Advances in Anatomy Embryology and Cell Biology

Vol. 139

Springer
Berlin
Heidelberg
New York
Barcelona
Budapest
Hong Kong
London
Milan
Paris
Santa Clara
Singapore
Tokyo

M. Elger T. Sakai W. Kriz

The Vascular Pole of the Renal Glomerulus of Rat

With 41 Figures

Springer

Dr. Marlies Elger
Institut für Anatomie und Zellbiologie
Universität Heidelberg
Im Neuenheimer Feld 307
69120 Heidelberg
Germany

Prof. Dr. Tatsuo Sakai
Department of Anatomy
Juntendo University
2-1-1 Hongo, Bunkyo-ku
Tokyo 113
Japan

Prof. Dr. Wilhelm Kriz
Institut für Anatomie und Zellbiologie
Universität Heidelberg
Im Neuenheimer Feld 307
69120 Heidelberg
Germany

ISBN-13: 978-3-540-63241-2 e-ISBN-13: 978-3-642-80449-6
DOI: 10.1007/978-3-642-80449-6

Library of Congress-Catalog-Card Number
Elger, M. (Marlies) The Vascular Pole of the Renal Glomerulus of rat / M. Elger, T. Sakai, and W. Kriz. p. cm. – (Advances in anatomy, embryology, and cell biology; 139) Includes bibliographical references and index.
ISBN-13: 978-3-540-63241-2
3. Rats-Physiology. I. Sakai, T. (Tatsuo), 1953- . II. Kriz, Wilhelm, 1936- . III. Title. IV. Series. QP249.E44 1997 573.4'9619352-dc21

Production: PRO EDIT GmbH, D-69126 Heidelberg
SPIN: 10570413 27/3136-5 4 3 2 1 0 Printed on acid-free paper

Acknowledgements

We appreciate the excellent technical assistance of Ms. Hiltraud Hosser and Ms. Bruni Hähnel, the skillful artwork by Mr. Rolf Nonnenmacher and the expert photographic work by Ms. Ingrid Ertel. We very much appreciate the valuable help received from Dr. Lise Bankir (Paris), Dr. Josie Briggs (Washington), Dr. Daniel Casellas (Montpellier), Dr. Armin Kurtz (Regensburg), Dr. Jürgen Schnermann (Ann Arbor) and Dr. Michael Steinhausen (Heidelberg). Dr. T. Sakai was a fellow of the Alexander von Humboldt Foundation. Financial support was provided by the Deutsche Forschungsgemeinschaft.

On the occasion of his 60th birthday, M.E. and T.S. wish to express their gratitude to Professor Dr. Wilhelm Kriz for constant guidance and exceptional scientific cooperation.

Contents

List of Abbreviations

AA	Afferent arteriole
EA	Efferent arteriole
– EA_{OS}	Outflow segment of EA
– – EA_{OSe}	portion of EA_{OS} in the EGM
– – EA_{OSi}	intraglomerular portion of EA_{OS}
– EA_{DS}	distal segment of EA in the cortical interstitium
EGM	Extraglomerular mesangium
BC	Bowman's capsule
– PBM	Parietal basement membrane
GBM	Glomerular basement membrane
JGA	Juxtaglomerular apparatus
MD	Macula densa
P_{AA}	Hydraulic pressure in the afferent arteriole
P_{EA}	Hydraulic pressure in the efferent arteriole
P_{GC}	Hydraulic pressure in glomerular capillaries
P_{M}	Hydraulic pressure in the mesangium
R_{AA}	Resistance of afferent arteriole
R_{EA}	Resistance of efferent arteriole
R_{OS}	Resistance of outflow segment

1 Introduction

Glomerular filtration represents one of the basic mechanisms in the function of an organism. Our understanding of this process is still quite fragmentary. Regulation of blood flow and pressure, together with regulation of the ultrafiltration coefficient (which is an attribute of the filtration barrier), are the two fundamental mechanisms accounting for maintenance and adaptability of glomerular filtration. Regulation of glomerular blood flow is generally considered to result from an interplay between afferent and efferent glomerular arterioles, and much progress has been made recently in understanding this interplay (Navar et al. 1996).

The present study provides a detailed structural description of the glomerular vascular pole of rat. The results of this study appear to be relevant for several open questions of glomerular function. First, the interaction between afferent and efferent arterioles in regulating glomerular blood is generally understood to occur between the preglomerular and the postglomerular portions of these vessels. As shown in the present study, the structural elaborations of these arterioles and the spatial relationships between them within the glomerular hilum strongly suggest an interplay also at this site. Moreover, the current understanding of glomerular blood flow regulation by tuning the interplay between afferent and efferent arterioles is exclusively based on signals whose regulatory loops are established in follow-up events outside the glomerulus (tubuloglomerular balance, tubuloglomerular feedback). The present study provides evidence that, in addition, regulatory loops may exist that directly connect the filtration gain to changes in glomerular hemodynamics; the structural organization of the glomerular hilum appears to provide such possibilities.

The results of this study are also relevant with respect to a better understanding of the support function of glomerular structures. The structural integrity of the glomerulus has to be maintained against the distending forces exerted on the glomerulus by the high hydraulic pressure in its center. At the vascular pole, the extraglomerular mesangium appears to be most important in serving as a closure device to hold the various structures at the vascular pole together (Kriz et al. 1988). In this function the extraglomerular mesangium (EGM) may be included in the tubuloglomerular feedback regulation of glomerular filtration.

The study deals with superficial and midcortical glomeruli. These glomeruli establish a homogeneous population, which is different from the glomeruli in the juxtamedullary region (Beeuwkes 1971; Kriz and Dieterich 1970). The presentation of the results will start with the description of the opening in Bowman's capsule, i.e., the glomerular hilum together with the structural organization of its walls, followed by the structures contained in the hilum, i.e., the glomerular arterioles, and is completed by the closure device of this opening, the EGM.

There is some overlap between the structures subsumed in the present study and the juxtaglomerular apparatus (JGA). Relationships of vascular pole structures to the macula densa have been extensively studied and have previously been reviewed (Barajas 1970; Barajas et al. 1990; Taugner and Hackenthal 1989). Therefore, the macula densa and its interface with the vascular pole were excluded from the description. We will touch on this problem in the discussion.

2 Material and Methods

Kidneys of Wistar rats (Ivanovas, Kissleg; 150–300 g BW) were fixed by total body perfusion as described previously (Kaissling and Kriz 1982). After anesthesia with thiobutabarbital (Inactin, 10 mg/100 g BW) or pentobarbital (Nembutal, 0.12 ml/100 g BW i.p.), the abdominal cavity was opened, the aorta clamped beneath the renal arteries, and a catheter was inserted into the abdominal aorta distal to the clamps in a retrograde direction. The catheter contained less than 0.5 ml of a solution prepared of 2 ml of 0.9% sodium chloride and 1 ml of heparin (5,000 IU). After the catheter was connected to the perfusion device, the inferior vena cava was broadly opened, the clamp was removed, and perfusion was started. Without prior flushing, perfusion of the fixative was carried out at a pressure of 180–220 mm Hg for 3 min at room temperature. The fixative contained 1.5% glutaraldehyde, 1.5% formaldehyde, 0.05% picric acid, and 0.08% $CaCl_2$ in 0.1 M cacodylate buffer (pH 7.4). Small pieces of cortical tissue were postfixed in buffered 1% OsO_4 dehydrated through a graded series of ethanol and embedded in Epon 812 by standard procedures.

In addition, a modified technique was used, resulting in improved preservation of intracellular filaments and extracellular matrix (Sakai and Kriz 1987). After perfusion fixation, tissue blocks of the same kidneys used above were only briefly (20 min) postfixed in a very low concentration of OsO_4 (0.1%) in 0.1 M cacodylate buffer. The blocks were then stained in the dark with 1% tannic acid (Mallinckrodt, Paris, KY) in 0.05 M maleate buffer at pH 6.0 for 3 h at 20 °C, followed by 1% uranyl acetate in the same buffer for 2 h at 4 °C. The tissue was dehydrated in a graded series of acetone solutions (1 h each) at gradually decreasing temperatures: 30% at 5 °C, 50% at 0 °C, 70% at –20 °C, 90% at –30 °C, and three baths of 100% at –30 °C (Carlemalm et al. 1982). Immersion into a 1:1 mixture of propylene oxide and Epon 812 was carried out at –30 °C for 24 h. The temperature was then allowed to rise, and infiltration with Epon and polymerization were accomplished by routine procedures. Semithin sections were stained with methylene blue B and azure II; ultrathin sections were stained with uranyl acetate and lead citrate and observed in a Philips 301 transmission electron microscope (TEM).

Superficial and midcortical glomeruli were exclusively investigated in this study. Series of ultrathin sections were prepared from entire glomerular vascular poles. The section planes were oriented either strictly in parallel or perpendicular to the longitudinal axis of the glomerulus, which resulted in longitudinal or cross sections through the vascular pole. These section series were used to examine the location and orientation of cells and cellular substructures (e.g., contractile proteins) and the overall spatial relationships of structures at the vascular pole. Furthermore, the findings were confirmed by observations made in numerous random sections of cortical tissue.

3 Results

The opening of Bowman's capsule allowing the entry and exit of glomerular arterioles is oval in shape (Fig. 1). The border of this opening is represented by the reflection of the glomerular visceral epithelium in the glomerular parietal epithelium. The opening contains both arterioles. The afferent arteriole (AA) has already lost its layer of smooth muscle; thus, only an endothelial tube enters the glomerular tuft. The efferent arteriole (EA) has a smaller lumen than the AA but, on the other hand, has a thicker wall that is established by an endothelial layer and at least one additional layer. Between the two arterioles is the waist of the opening. At this site the extraglomerular mesangium (EGM) passes over into the mesangium proper; this transition is established by only a few cells.

Strictly speaking, the geometrical vascular pole of the glomerular sphere may be considered as a point that is located within the EGM between the glomerular arterioles. We define two meridians through the geometrical vascular pole to facilitate description (Figs. 2, 3): a *transarteriolar meridian*, defined as extending through the

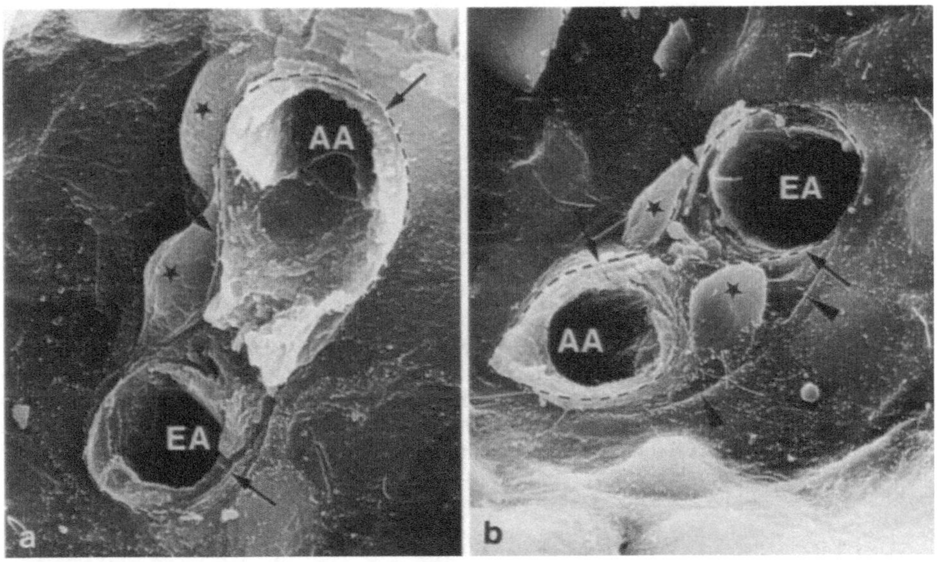

Figs. 1a,b. Vascular poles as seen by scanning electron microscopy (SEM) from inside of Bowman's capsule. The glomerular tufts are broken away; the edge of break is indicated by a *dotted line*. The enclosed area may be regarded as the glomerular stalk containing the afferent (*AA*) and efferent arteriole (*EA*). Outside the break line the edge of the parietal epithelium is seen (*arrowhead* in **b**); compare Fig. 5. Protruding cell bodies (*stars*) surrounding the vascular pole represent transitional cells exhibiting features of parietal cells or of incompletely differentiated podocytes. **a** ×~2000; **b** ×~1850. **b** Reprinted from Frank and Kriz (1982) with permission

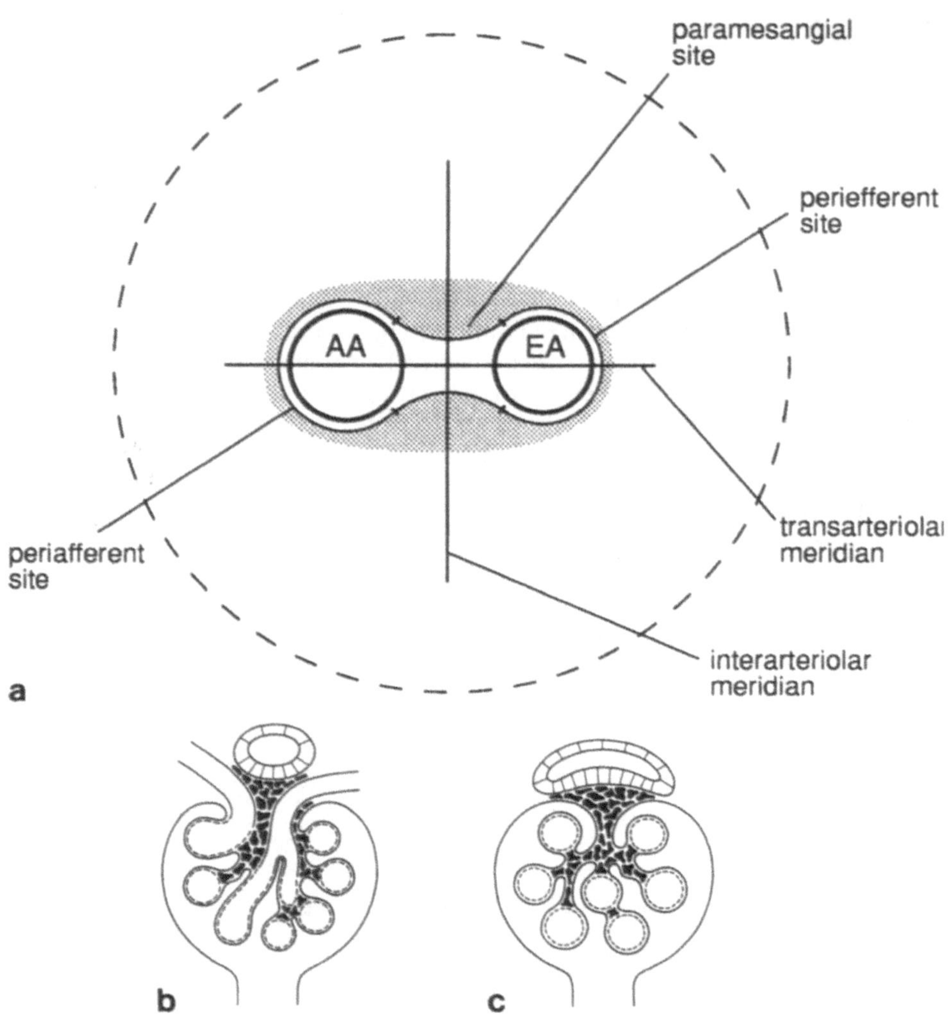

Fig. 2a–c. Schematics to define the nomenclature. The schematic in **a** shows an idealized view of the glomerular vascular pole from outside after removal of the juxtaglomerular apparatus. The gray area represents that part of the parietal basement membrane of Bowman's capsule, which serves for affixation of the extraglomerular mesangial cells and of arteriolar muscle cells. In this area a *paramesangial site* is distinguished from a *periarteriolar site*, which is subdivided into *periafferent* and *periefferent sites*. **b,c** Sectional profiles of glomeruli through the *transarteriolar meridian* and *interarteriolar meridian*, respectively

Figs. 3a,b. Meridional sections through the vascular pole of superficial glomeruli. **a** Transarteriolar section with afferent and efferent arteriole (*AA, EA*). Inbetween the two arterioles one can see the extraglomerular mesangium (*EGM*), which is in contact with the macula densa (*MD*). Towards the glomerular tuft the EGM merges into the glomerular mesangium proper (*M*). After entering the glomerulus, the AA immediately transforms into capillaries (*arrowhead*). In contrast, the EA emerges from the glomerular tuft as a vessel with a multilayered wall (*asterisk*). The border of the opening in

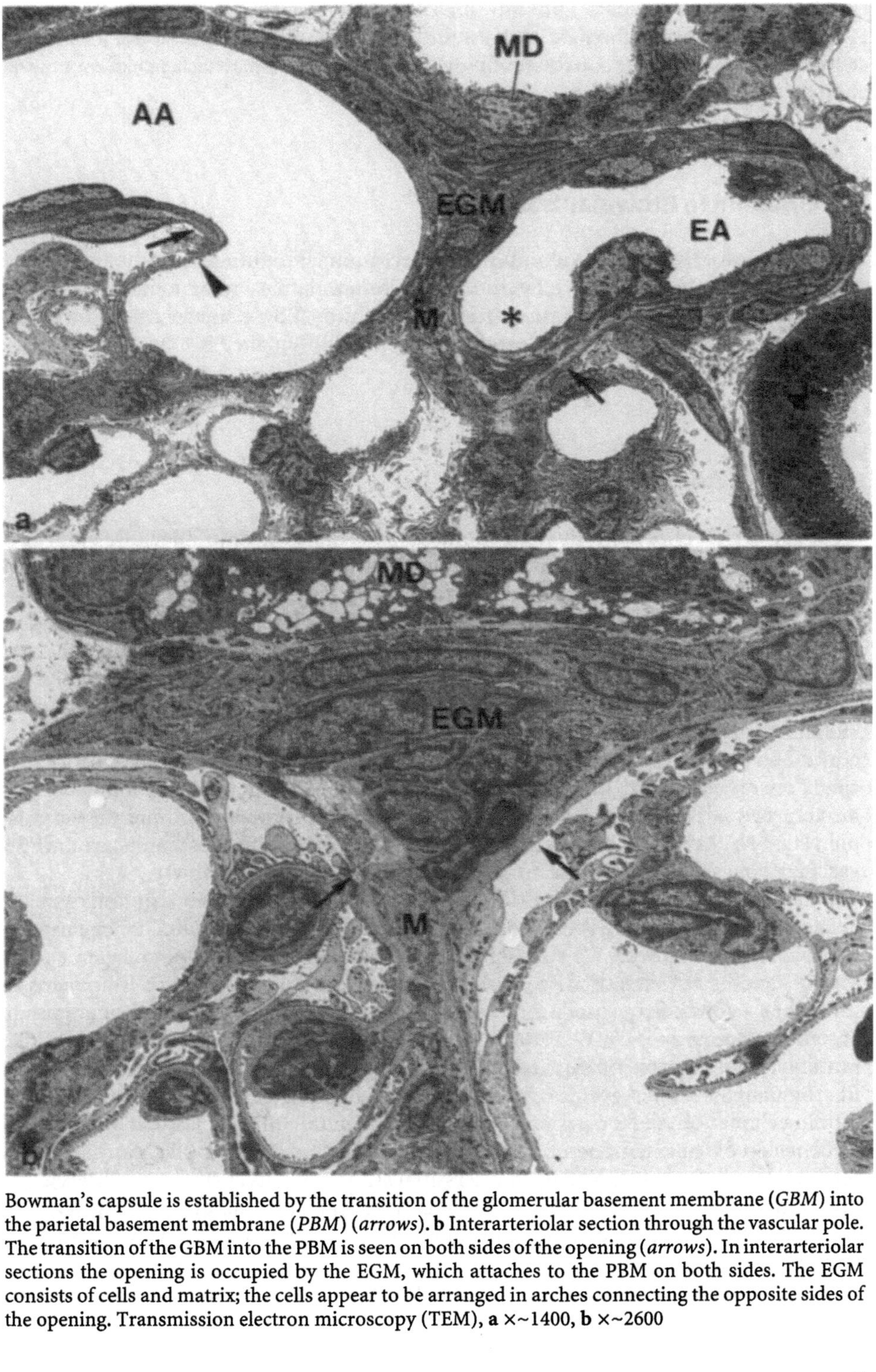

Bowman's capsule is established by the transition of the glomerular basement membrane (*GBM*) into the parietal basement membrane (*PBM*) (*arrows*). **b** Interarteriolar section through the vascular pole. The transition of the GBM into the PBM is seen on both sides of the opening (*arrows*). In interarteriolar sections the opening is occupied by the EGM, which attaches to the PBM on both sides. The EGM consists of cells and matrix; the cells appear to be arranged in arches connecting the opposite sides of the opening. Transmission electron microscopy (TEM), **a** ×~1400, **b** ×~2600

center of both arterioles, and an *interarteriolar meridian,* defined as running perpendicularly to the former, in the middle between the two arterioles. The region surrounding the opening can be subdivided into a *periafferent area,* a *periefferent area,* and two *paramesangial areas.*

3.1 The Opening in Bowman's Capsule

The reflection of Bowman's visceral epithelium (podocyte epithelium) into the parietal epithelium, together with the reflection of the glomerular basement membrane (GBM) in the basement membrane of the parietal epithelium (PBM), borders the glomerular entrance. Strictly speaking, it is the reflection of the GBM in the PBM that actually faces this opening.

3.1.1 Transition of the GBM into the PBM

At the epithelial reflection, the thick homogeneous dense layer of the GBM splits into several thin, dense layers, establishing the multilayered PBM. The multiple dense layers are separated from each other by translucent layers of varying thickness (Figs. 4, 5a–c). The dense layers of the PBM are basically composed of fine filaments, ranging from 5 to 15 nm in diameter. They assemble to form filament bundles up to 100 nm thick and 1–2 µm in length. Within a dense layer, filaments and filament bundles are oriented mainly in preferential directions varying from layer to layer (Mbassa et al. 1988). At its inner aspect, adjacent to the epithelium, the PBM often does not form continuous layers, but the filament bundles are arranged in ribbonlike structures called *"microligaments"* (Mbassa et al. 1988). These microligaments are embedded in basal furrows of the parietal epithelium and are most prominent around the vascular pole (Fig. 5b). At this site they appear to form nearly continuous bands around the vascular pole associated with actin bundles inside the cell (see below).

In addition to type 4 collagen (Butkowski et al. 1989), the PBM contains type 14 collagen belonging to the FACIT collagen subfamily (fibril-associated collagens with interrupted triple helices) (Lethias et al. 1994). Usually, this type of collagen is seen around bundles of interstitial collagen fibrils (e.g., in tendon or skin). It appears to function as a connecting medium between proteoglycans and other collagen types of basement membranes. In the PBM it might be relevant for the establishment of the multilayered structure of this basement membrane as well as for formation of the microligaments. It has been proposed that type 14 collagen could act as a spacer or binding element in the formation of fibril bundles in the PBM. This could be achieved by cohesion of basement membrane units of Bowman's capsule in a fashion similar to that postulated for this molecule in interstitial striated fibrils (Lethias et al. 1994). The predominant proteoglycan of the PBM is a chondroitin sulfate proteoglycan, in contrast to the dominance of heparan sulfate proteoglycan in the GBM (Couchman et al. 1994).

At its outer aspect, especially around the vascular pole, the PBM is regularly accompanied by *reticular fibrils.* They are easily identified by their typical thickness

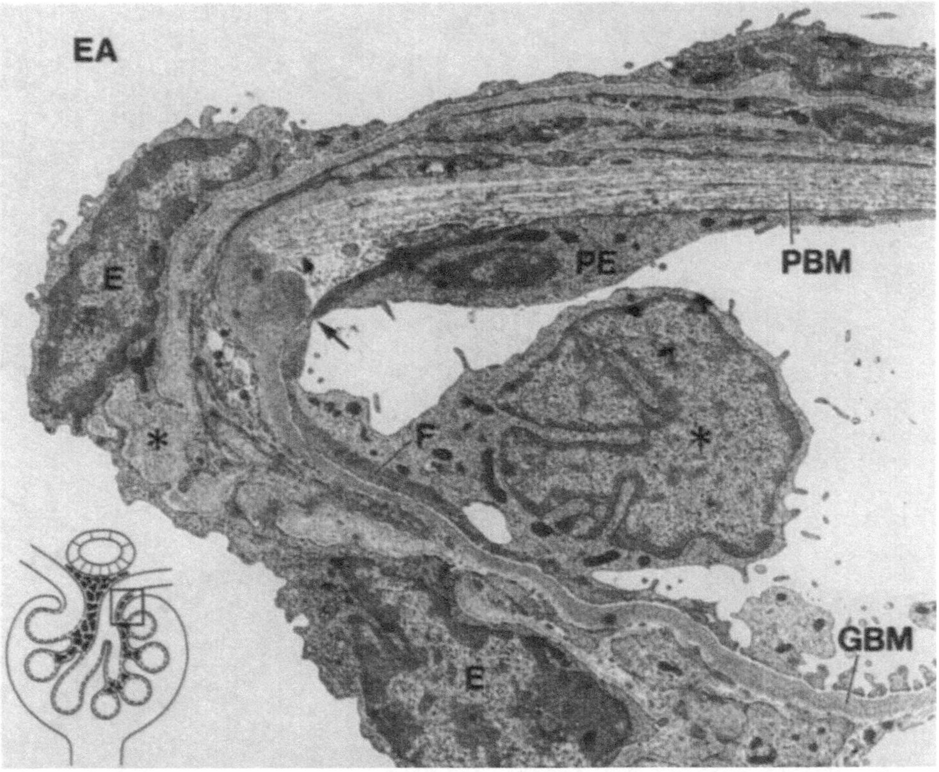

Fig. 4. Transarteriolar section through the efferent arteriole (*EA*), showing the transitions of visceral epithelium/GBM into parietal epithelium (*PE*) of Bowman's capsule/parietal basement membrane (*PBM*). At the transition the compact GBM splits into the multilayered PBM (*arrow*). The outermost podocyte (*asterisk*) attaches to the GBM without forming foot processes; the basal cytoplasm contains a layer of densely packed microfilaments (*F*). The capillary endothelium (*E*) and the GBM are separated by matrix (*asterisk*), which continues into the subendothelial space of the EA. TEM, ×~6100

(about 30 nm) and cross-striped pattern (Fig. 6); they contain type III collagen. No specific orientation of these fibrils is apparent; they seem to run in all directions.

Again most prominent at the vascular pole, accumulations of *membrane-bound, granular vesicles* are often found to be associated with the PBM (Figs. 5c,6). Generally, they are located between the outermost dense layers of the PBM. Such vesicles, which have also been found in rat and human glomeruli, are called "spherical microparticles" (Burkholder et al. 1973), "granular and vesicular structures" (Johnston et al. 1973), or "membrane-limited particles" (Olsen et al. 1974). The diameter of these vesicles ranges from 25 to 100 nm, but also larger vesicles up to 500 nm are encountered. They contain material of varying density (hyaline, electron-dense, lamellated) and frequently cellular remnants such as fragments of mitochondria or endoplasmic reticulum. They appear to represent cell debris left over from cell disintegration.

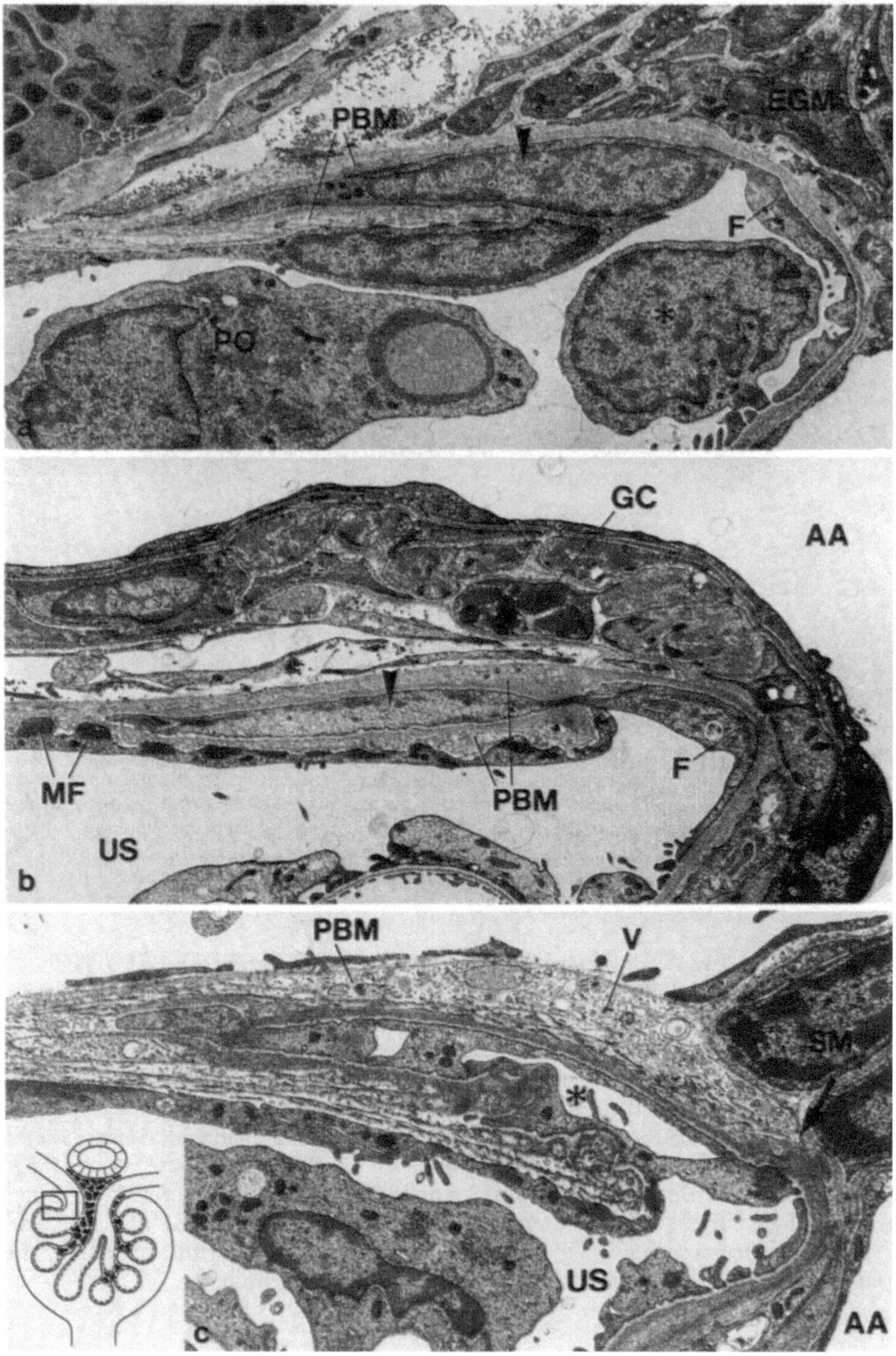
PBM
EGM
F
PO
GC
AA
MF
PBM
F
US
b
PBM
V
SM
US
AA
c

3.1.2
Transition from Podocytes to Parietal Cells

The transition of the visceral epithelium (podocytes) into the flat parietal epithelium is fairly abrupt and is located around the stalk of the glomerular tuft. The last row of podocytes and the first parietal cells at this transition frequently exhibit specific structural features.

The *last podocytes* have few cell processes and contain microfilament bundles or belts, characterized by serial densities typical of stress fibers (Figs. 4, 7). They are reminiscent of cytoskeletal changes in podocyte architecture, going along with foot process effacement in pathological situations (Shirato et al. 1996).

The first row (or rows) of the *parietal cells* exhibits several structural features obviously accounting for an intimate association with the PBM. First, the parietal cells, at their basal aspect, generally possess *cytoplasmic ridges* containing prominent *bundles of microfilaments* (actin filaments) (Pease 1968; Mbassa et al. 1988). At the vascular pole, these ridges appear more prominent than in other portions of Bowman's capsule, consisting of up to 12 parallel ridges. Individual ridges have a height of up to 600 nm containing microfilament bundles of 500 nm thickness. As seen in appropriate sections, these ridges run along the opening of Bowman's capsule (Figs. 5b, 8; see also Fig. 20b). They are more numerous and appear more extensively developed on the perimesangial than on the periarteriolar side (seen in 70% versus 50% of random longitudinal sections of the vascular pole). The extracellular "microligaments" of the PBM run in parallel to the microfilament bundles. Both are connected to each other by focal adhesions containing the extracellular adhesive protein fibronectin (unpublished results from our laboratory). Thus, two supporting structures are arranged in parallel "rings" around the vascular pole: the actin filament bundles within the basal ridges of the parietal cells and the extracellular microligaments within parallel furrows (Fig. 9).

Second, *outpocketings* of the parietal epithelium are frequently found at the vascular pole. These outpocketings – variable in extension and depth – invaginate into the PBM. They may be especially prominent on the perimesangial sides. In conjunction with these outpocketings, recesses of the urinary space are established (Fig. 5c). In addition, parietal cells frequently form winglike processes protruding – like the outpocketings – into the PBM (Fig. 5b). They are located between the dense layers of the PBM. Often these processes are of considerable size; in extreme cases the cell body together with the nucleus is included within the PBM (Fig. 5a). Most frequent are cytoplasmic processes of smaller size; small projections may also extend backwards toward the

Figs. 5a–c. Meridional sections through the afferent arteriole (*AA*), showing the variable organization of the parietal epithelium at the transition. **a** The cell body of a parietal cell (*arrowhead*) projects deeply between the layers of the PBM above the parietal epithelium. **b** Also, in this case, one of the parietal cell (PE) forms a winglike process (*arrowhead*), which is embedded in the PBM. The PE facing the urinary space (*US*) contains prominent basal ridges filled with microfilament bundles (*MF*). A granular cell (*GC*) is seen in the wall of the AA. *F*, microfilament bundles at the reflection of the epithelium. **c** In this case, a recess (*asterisk*) of the US is found, which is bordered by parietal cells. The GBM transforms abruptly into the multilayered PBM (*arrow*). Inside the PBM many vesicles (*V*) are seen. Note the attachment of an arteriolar smooth muscle cell (*SM*) to the PBM. *PO*, podocyte. TEM, **a** ×~6500, **b** ×~6100, **c** ×~7650

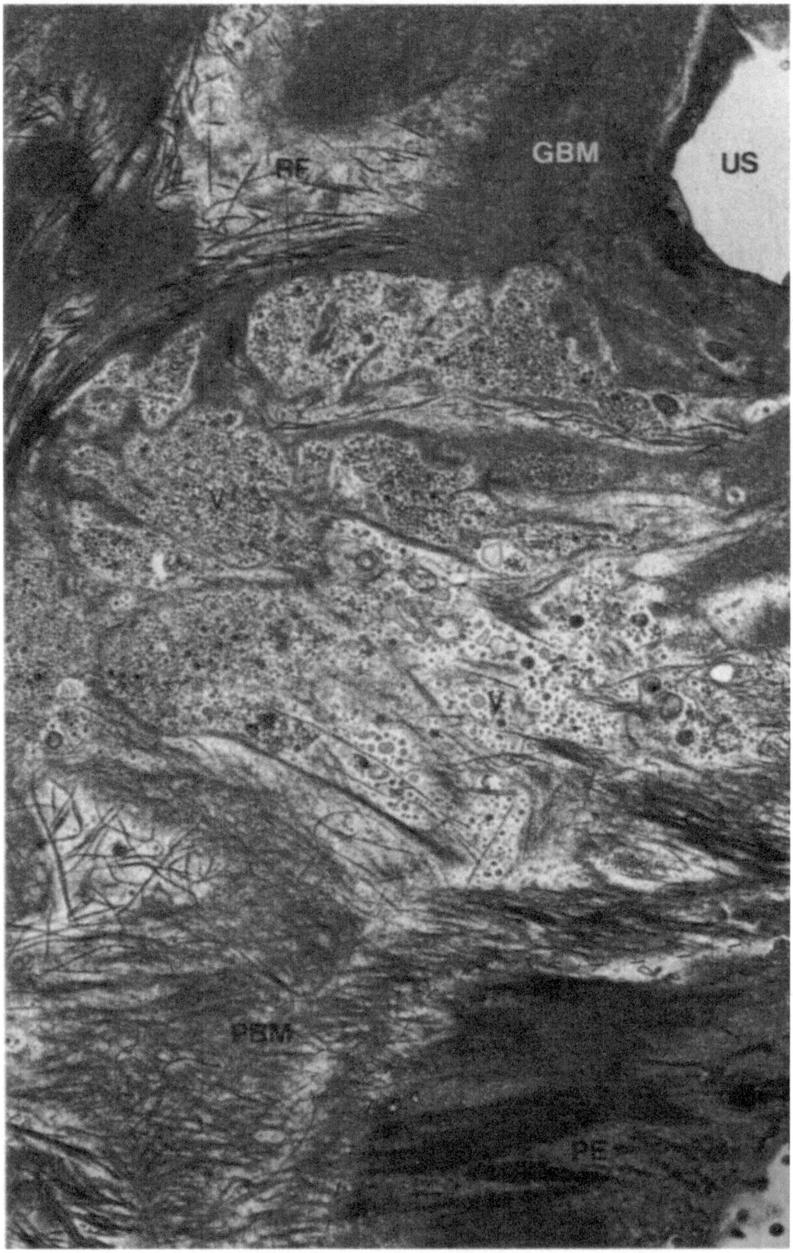

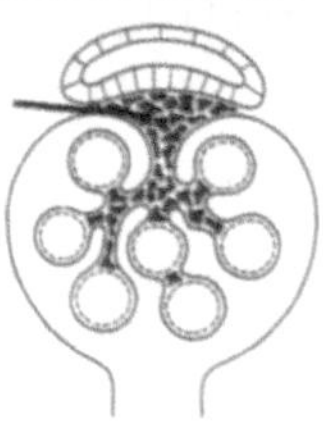

Fig. 6. Grazing section through the transition of the GBM to the PBM. *At the top* the homogeneous dense matrix of the GBM is seen facing the urinary space (*US*). A parietal cell (*PE*) is cut at the bottom. The middle part of the figure, delineated by a *broken line*, represents the PBM, consisting of dense layers and of expanded spaces filled with vesicles (*V*). Note the many reticular fibers (*RF*) at the interface to the cortical interstitium. TEM, ×~12 300

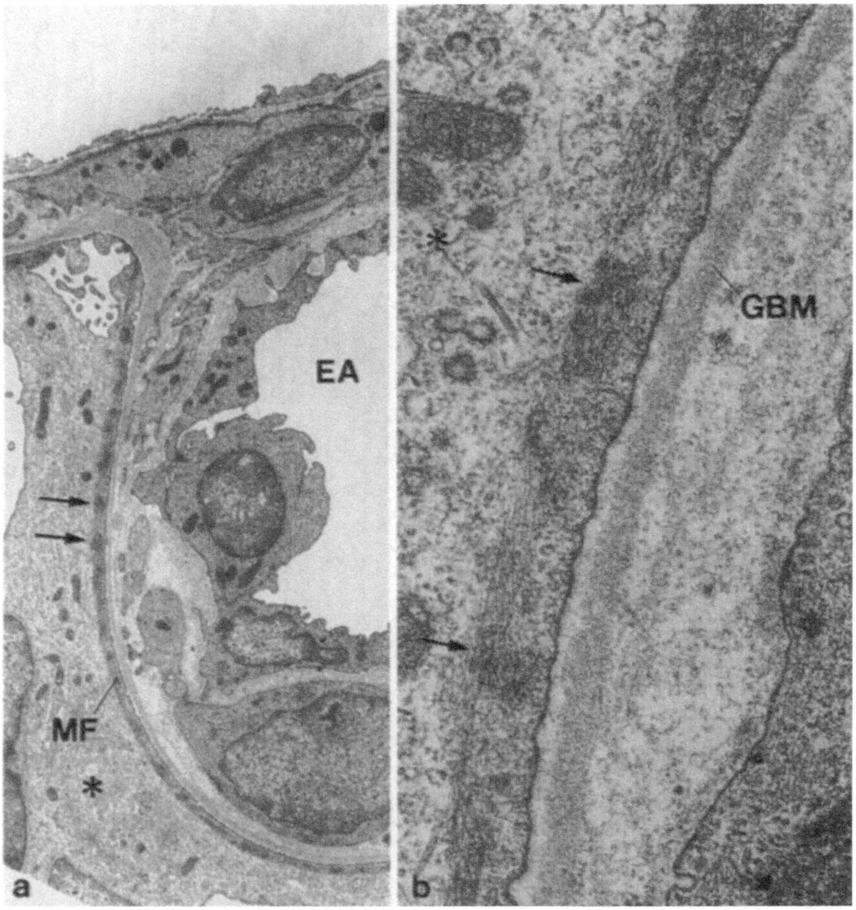

Figs. 7a, b. Cross section through the intraglomerular portion of the efferent arteriole (*EA*) showing details of the cytoskeleton at the transition of the epithelium. The last podocyte (*asterisk*) before the transition into the parietal epithelium is attached to the GBM without forming foot processes (compare Fig. 4). The basal cytoplasm contains a band of microfilaments (*MF*) running in a circular direction around the EA. Periodic densities (*arrows*) of the bundle are seen at a higher magnification (**b**). (Parallel section of section in Fig. 19a). TEM, **a** ×~6400, **b** ×~47 500

glomerular stalk. These cell processes and outpocketings appear to be stretched out within the PBM in meridional directions.

As a result of these outpocketings and cell processes the parietal epithelium forms edges protruding into the urinary space (Figs. 5b,c). These edges run partly or entirely around the two arterioles and are frequently apparent in scanning electron micrographs showing the inner surface of Bowman's capsule (Fig. 1). In contrast to the outline of the opening of Bowman's capsule, which displays a waistlike indentation, this edge is nearly oval in shape and surrounds the glomerular entrance on its inner side. Interestingly, also the major microfilament bundles inside these cells (described above) follow this oval line around the vascular pole (Fig. 8; see also Fig. 20b).

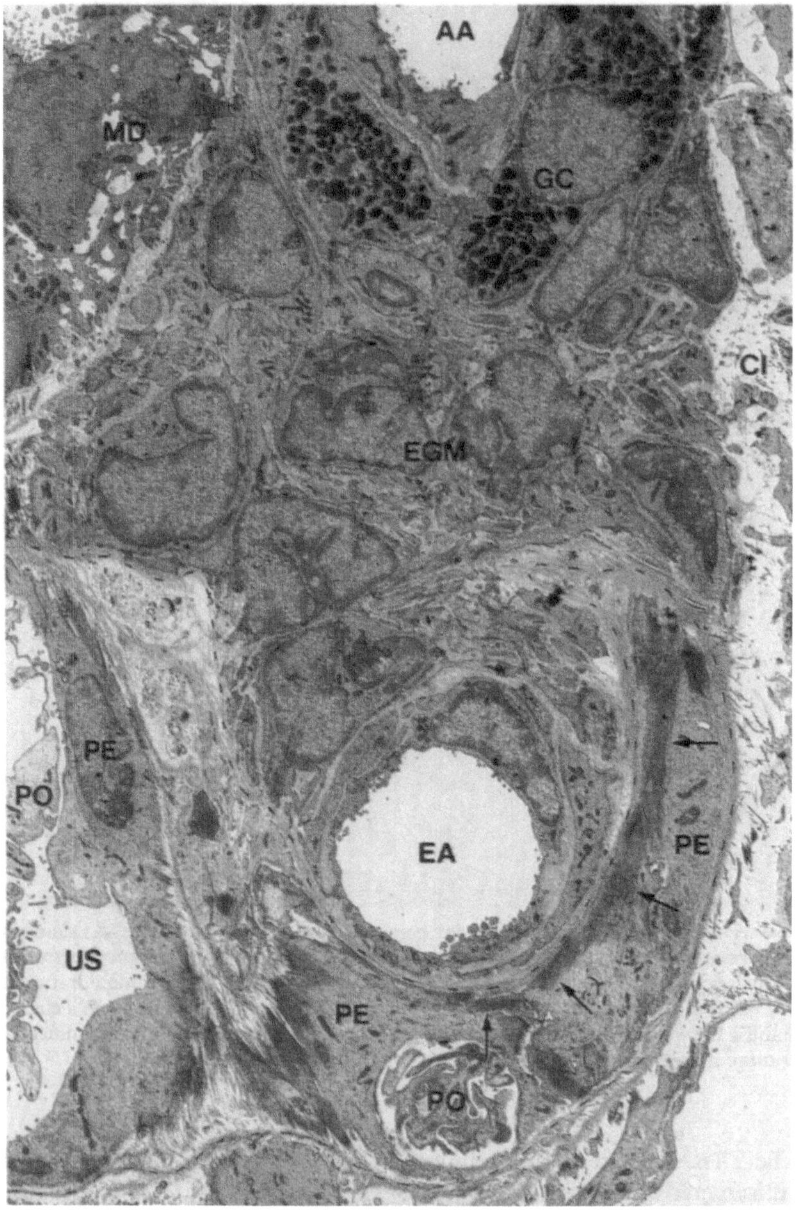

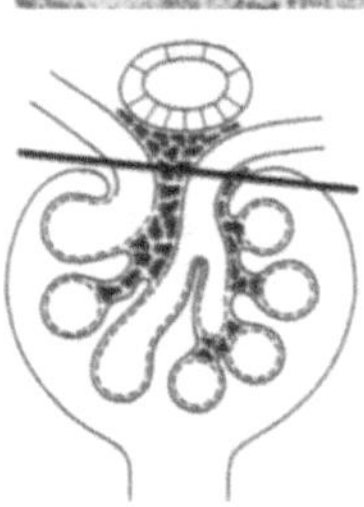

Fig. 8. Cross section through the vascular pole. Between afferent (*AA*) and efferent (*EA*) arterioles the extraglomerular mesangium (*EGM*) is seen. In parietal cells (PE, together with the PBM delineated by a *stippled line*) prominent microfilament bundles (*arrows*) tend to encircle the efferent arteriole. *CI*, cortical interstitium; *MD*, macula densa; *PO*, podocyte; *US*, urinary space. TEM, ×~3300

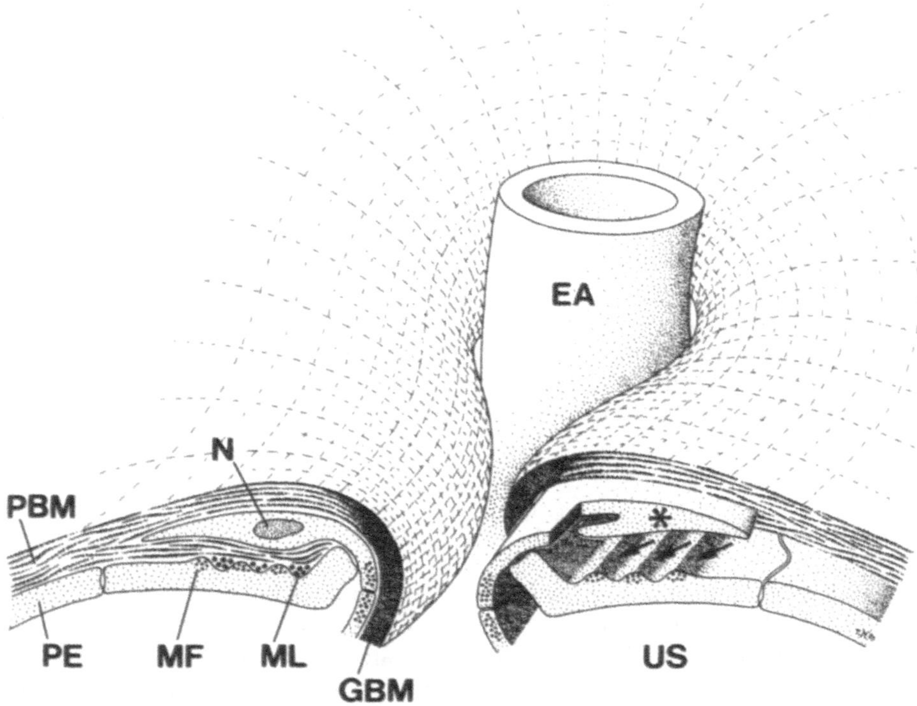

Fig. 9. Schematic to demonstrate the wall structure of Bowman's capsule at the glomerular entrance. An interarteriolar section with a survey view of one-half of the glomerular entrance containing the efferent arteriole (*EA*) is shown. Mesangial and extraglomerular mesangial cells are not drawn. The outermost layer seen from above is the transition of the multilayered parietal basement membrane (*PBM*) into the homogeneous GBM. On the *right side*, the basement membrane is partly removed to show the cellular organization at the transition. The parietal cells (*PE*) at this site exhibit several peculiarities: first, they may form winglike processes, which run in a meridional direction between the layers of the PBM (*asterisk*). A nucleus (*N*) of a PE in this location is marked. Second, the PE facing the urinary space (*US*) have prominent basal ridges (*arrows*) that contain circularly running microfilament bundles. The extracellular furrows between these ridges are filled with collagenous microligaments (*ML*) of the PBM running in the same direction. Also, the cell portions of PE or of podocytes at the bend of the epithelium contain microfilament bundles (*MF*) running in a circular direction

A further type of specialization of the parietal epithelium at the vascular pole has to be considered, the so-called *peripolar cells*, originally described in sheep by Ryan et al. (1979). They found that in sheep the parietal cells at the vascular pole may contain abundant cytoplasmic granules reminiscent of renin granules in granular cells of the afferent arteriole. Subsequent studies have shown that these granules contain a neuron-specific enolaselike protein (Trahair and Ryan 1989) and transthyretin (Hollywell et al. 1992). The number of cells and the number of granules per cell vary greatly among species and, furthermore, are dependent on age (Gall et al. 1986; Mbassa 1991). In the rat kidney, granulated peripolar cells have only been found very rarely

(Gall et al. 1986; Downie et al. 1992). In all our investigations we ourselves only observed a single cell.

In the human, podocytes are encountered that are included in the parietal epithelium at the vascular pole ("parietal podocyte"; Gibson et al. 1992, 1996). The cell bodies of these podocytes rest on the PBM. With neighboring cells they form interdigitations (by foot processes) and slit membranes, and they may send processes onto glomerular capillaries. In rat corresponding cells were only occasionally found, whereas cells with voluminous cell bodies protruding into the urinary space, but without significant formation of processes, were a common feature (Fig. 1; see also Figs. 15 and 26). Whether the latter represent a transitional cell type between podocytes and parietal cells, incompletely differentiated podocytes or simply a structural variation of parietal cells is unknown. None of these cell types is regularly found. Other investigators consider any prominent cell at this site, protruding into the urinary space, as a kind of (active or inactive) peripolar cell (Downie et al. 1992). As long as the peripolar cells cannot be defined functionally, this hypothesis cannot be tested.

3.2 Glomerular Arterioles

Immediately after its entry into the glomerular tuft, the AA splits into several superficially located branches. In contrast, the EA is already established deep within the glomerular tuft by the confluence of centrally located tributaries. These findings have been repeatedly shown by scanning electron-microscopic (SEM) studies in the rat (Murakami 1972; Spinelli 1974), in vivo studies in the hydronephrotic rat kidney (Steinhausen et al. 1983; Snoei 1987), and transmission electron-microscopic (TEM) studies in the rat (Kriz et al. 1988; Elger et al. 1991) and pig (Moore et al. 1992). The present study extends our knowledge about, first of all, the branching and confluence pattern of the arterioles, second, the spatial relationships between afferent and efferent vessels and, finally, the wall structure of these vessels at the vascular pole. It is necessary to mention again that the present study deals only with the arterioles of superficial and midcortical glomeruli; the arterioles of juxtamedullary glomeruli are different in several respects.

The branching and confluence pattern of glomerular arterioles is correlated with the subdivision of the glomerular tuft into lobules (usually 3, range 1–5) (Fig. 10). Each lobule has an afferent and an efferent domain. In the afferent domain the capillaries run towards the urinary pole, in the efferent domain towards the vascular pole. As a whole, the capillaries have an "U"-shaped form in each lobule. The afferent domains of all lobules together comprise about three-fourths to two-thirds of the tuft. The efferent domains are generally located together. establishing their own sector of the tuft. Towards the vascular pole, this efferent sector is, for the major part, surrounded by the afferent domains (Figs. 11, 12).

Adjacent lobules may be interconnected by few, if any, capillary bridges. No anastomoses between the first branches of the afferent domain and terminal vessels of the efferent domain, which could function as short pathways (shunt vessels), were found during reconstruction studies (Winkler et al. 1991), in corrosion casts (Spinelli et al. 1972; Murakami 1972; Unehira 1981; Moore et al. 1992), or intravital microscopy (Snoei 1987).

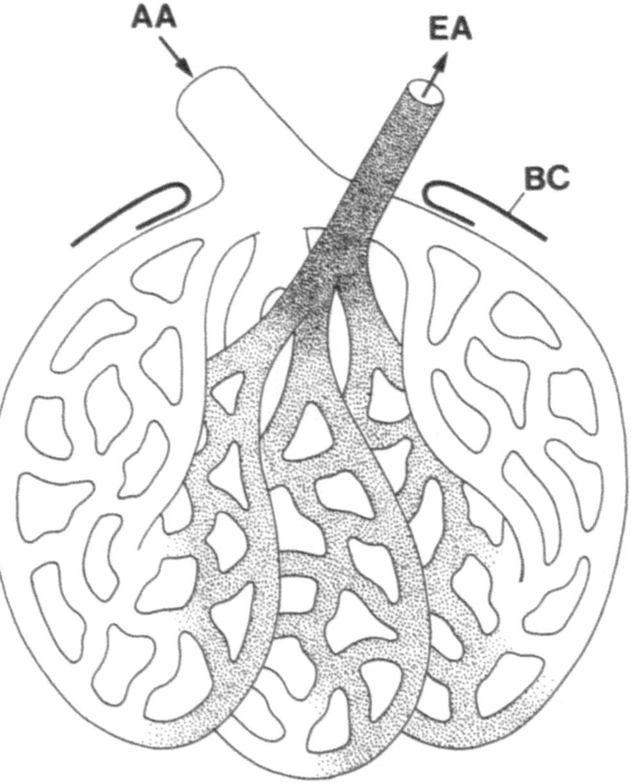

Fig. 10. Schematic of the glomerular microvasculature. Immediately after its entry into the tuft the afferent arteriole (*AA*) splits into several superficially located primary branches, each of which supplies one lobule (three are shown). The proximal portions of the lobules establish the afferent domain (shown *in white*), the capillaries of which, as a whole, run towards the urinary pole. The distal portions represent the efferent domain (*dotted texture*). The efferent arteriole (*EA*) is formed by the confluence of several tributaries within the efferent domain. Compared to the AA and its primary branches, the EA and its immediate tributaries have relatively narrow lumina. [From Winkler et al. (1991) with permission]

The EA leaves the glomerulus in close proximity to the AA (Figs. 3a, 11). The narrow region between the two arterioles is occupied by a few mesangial cells that establish the transition of the mesangium proper to the extraglomerular mesangium (EGM). Together with the surrounding GBM, these structures represent the *glomerular stalk* (see below).

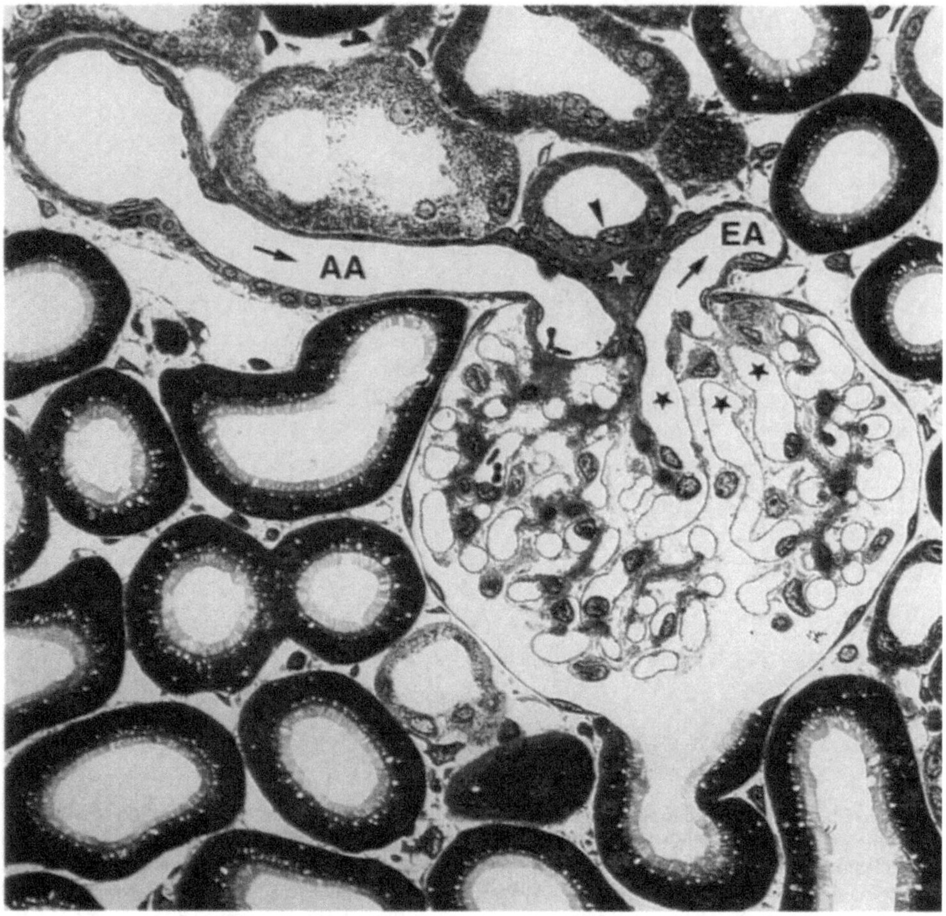

Fig. 11. Transarteriolar section through a glomerulus, overview. The afferent arteriole (*AA*) originates from a cortical radial artery and runs to the vascular pole. The efferent arteriole (*EA*) is established inside the glomerular tuft. Between the two arterioles the extraglomerular mesangium (*star*) and the macula densa (*arrowhead*) are seen. The *small asterisks* mark the tributaries to the EA. Note the tributary that joins the EA shortly before the latter leaves the glomerulus. Semithin Epon section; LM, ×~430

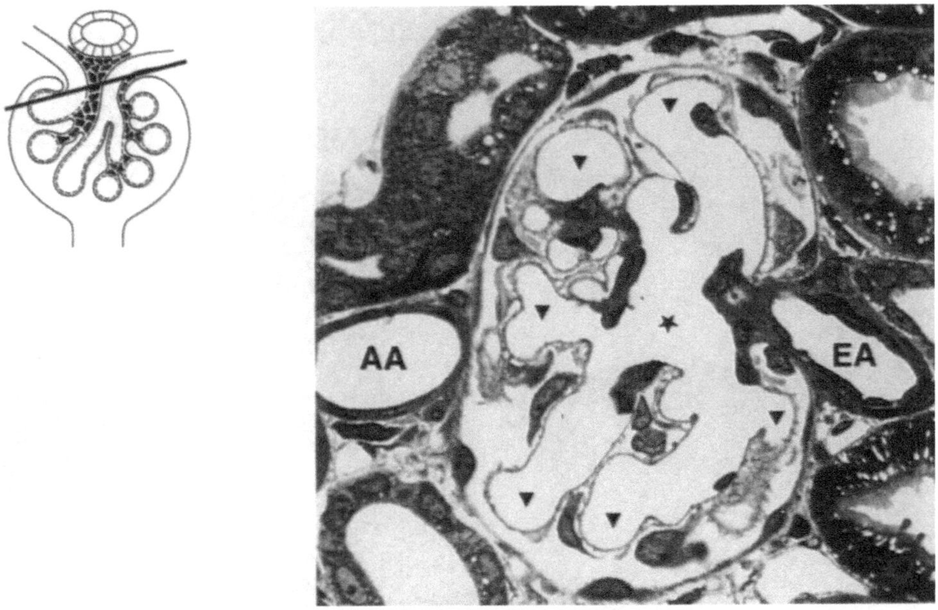

Fig. 12. Branching pattern of the afferent arteriole (*AA*). Cross section just beneath the vascular pole. The AA and EA are seen outside the glomerulus. At its branching point (*star*) inside the glomerulus the AA splits into large primary capillaries (*black triangles*), distributed on the surface of the tuft. Semithin Epon section, LM, ×~590

3.2.1 Afferent Arteriole

The wall of the AA consists of a thin, continuous endothelium, a basement membrane, and a single layer of smooth muscle cells (Figs. 13, 14). An internal elastic layer generally disappears some distance from the glomerulus. Near the glomerulus, smooth muscle cells are usually replaced by granular cells. These cells establish contact with the EGM. The smooth muscle layer is first lost on the side towards the center of the EGM (Figs. 15a; see also Fig. 29). On the opposite side, the muscle layer is maintained until the AA attaches to the parietal basement membrane (Fig. 14, 15a). As a consequence, the AA proper does not enter the tuft, but divides into its branches at the entrance level.

The AA is outlined by a continuous endothelium. The flat cells are interconnected by gap junctions (Taugner et al. 1978). Upon entering the glomerulus, the endothelium changes into the fenestrated type typical of glomerular capillaries (Fig. 13). Fenestrations first appear on the side towards the EGM (Fig. 15b). The basement membrane of the AA merges with the extracellular matrix of the mesangium and does not continue into the glomerulus.

The primary branches of the AA distribute on the surface of the glomerular tuft. Thus, most of the superficial vessels at the vascular pole belong to the afferent domain (Fig. 12). The primary branches have wider lumina than the bulk of glomerular

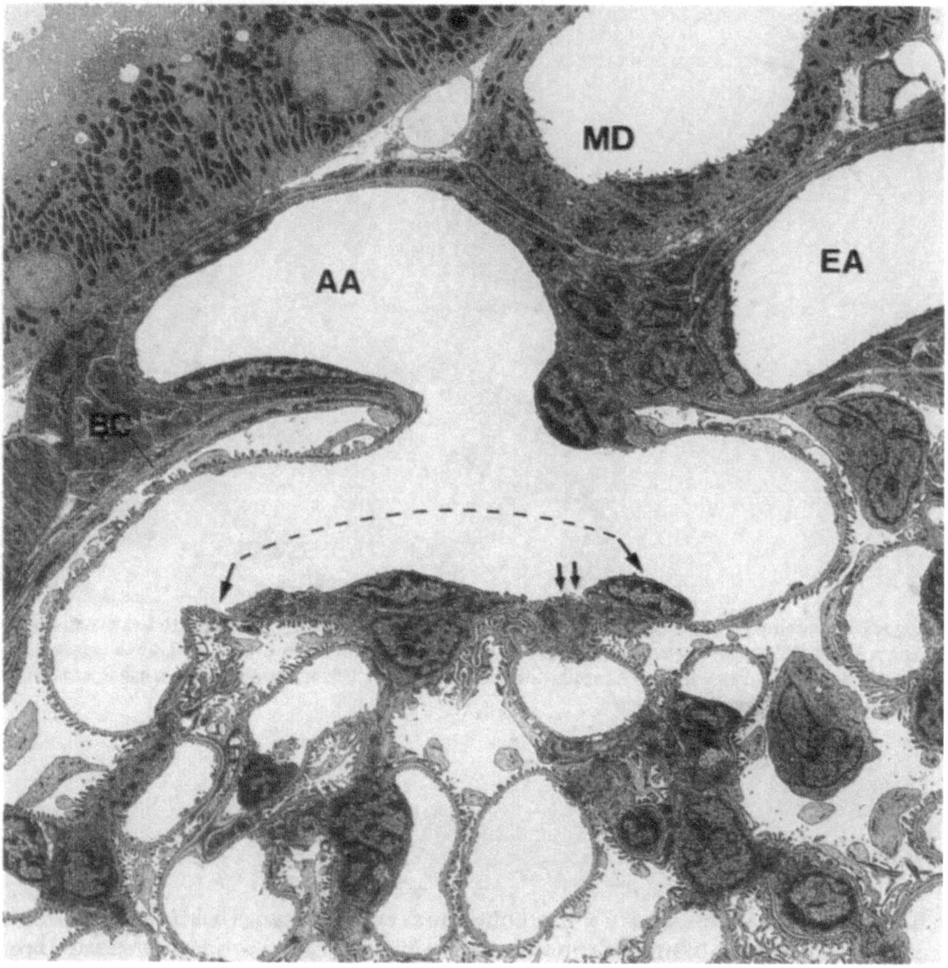

Fig. 13. Afferent arteriole. Longitudinal section through the vascular pole showing the entry of the AA into the glomerulus and its splitting into primary branches. These already have the characteristic structure of glomerular capillaries. Note the extensive contact zone (indicated by the *stippled line*) of the branching area of the AA to the mesangium; the endothelium at this site is fenestrated (*double arrow*). *BC*, parietal layer of Bowman's capsule. The macula densa (*MD*) is cut in its peripheral part. TEM, ×~1800

Fig. 14. Afferent arteriole at the entrance of the glomerulus. The peripheral wall of he AA faces the cortical interstitium and adheres to the transition between GBM and PBM. In addition to smooth muscle cells (*SM*) the wall contains granular cells (*GC*). The endothelium (*E*) of the AA already displays fenestrae. Immediately after the entry of the AA into the glomerulus the fenestrated endothelium is fully established. The transition from the parietal (*PE*) to the visceral epithelium with its interdigitating foot processes (*FP*) is rather abrupt. *C*, glomerular capillary; *IE*, internal elastica; *NE*, unmyelinated nerve; *PBM*, parietal basement membrane. TEM, ×~5000

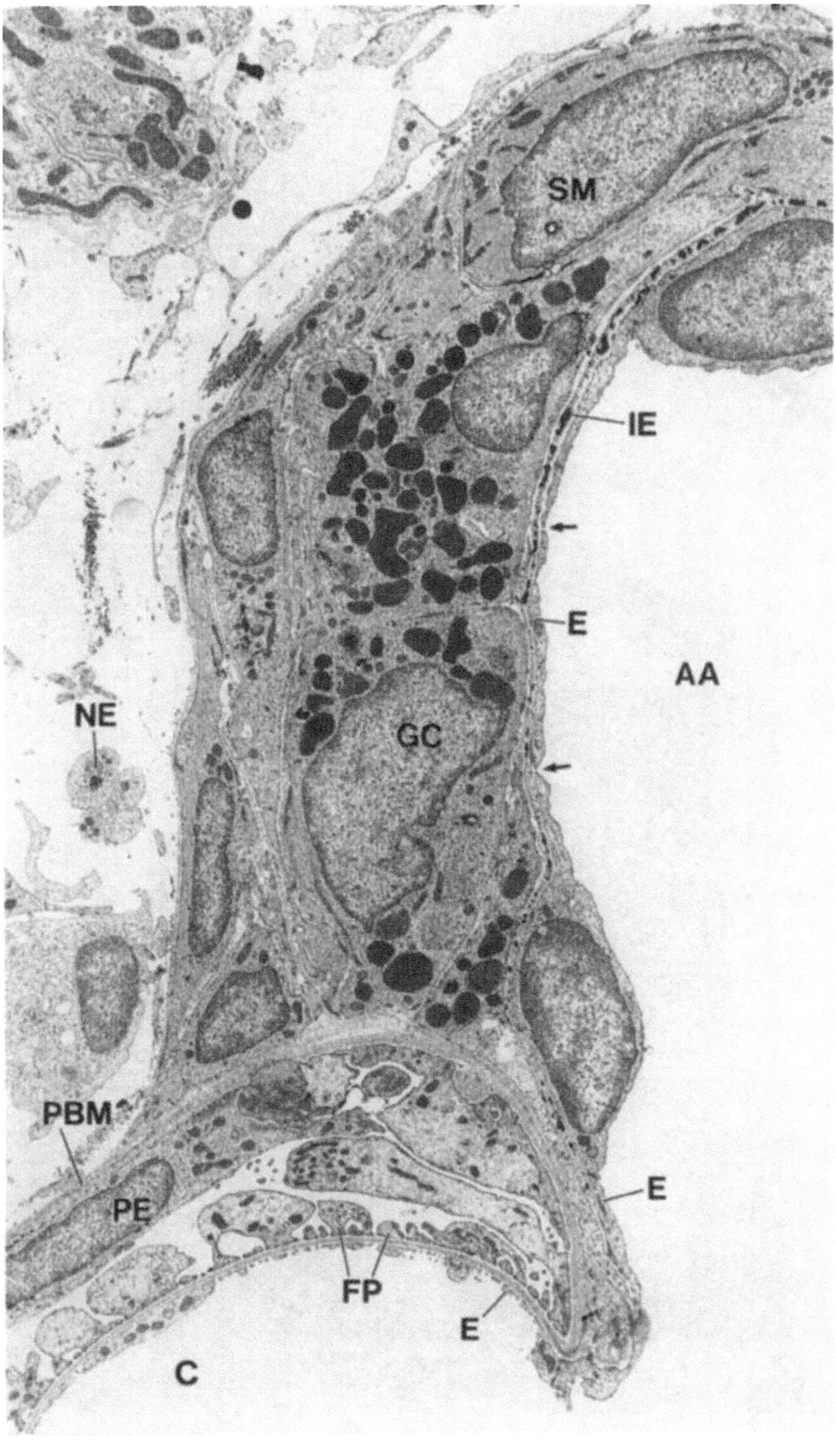
SM
IE
E
AA
NE
GC
PBM
PE
E
FP
E
C

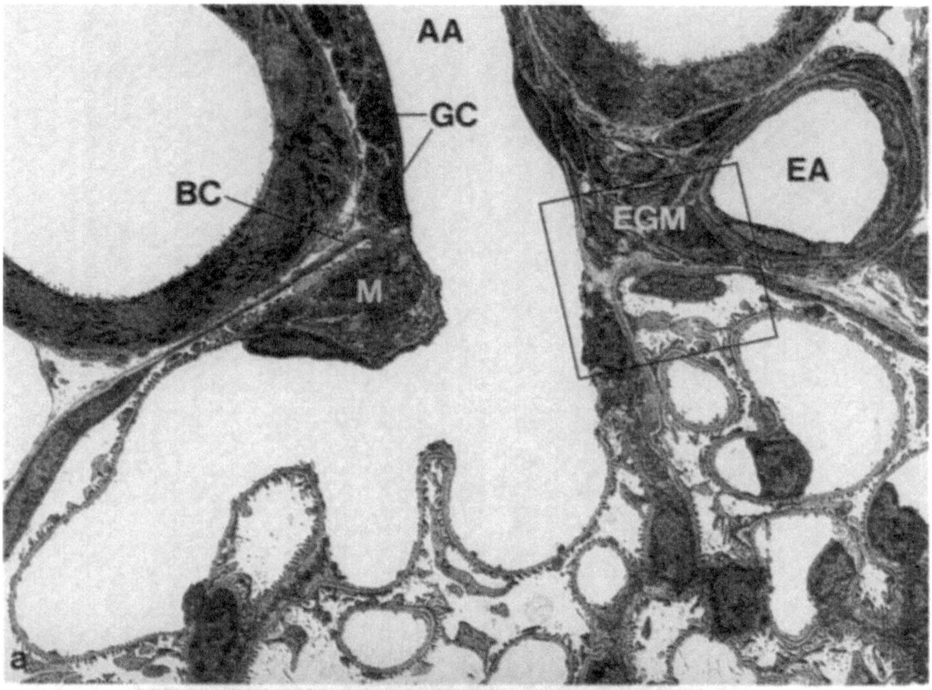
AA
GC
BC
M
EGM
EA
a

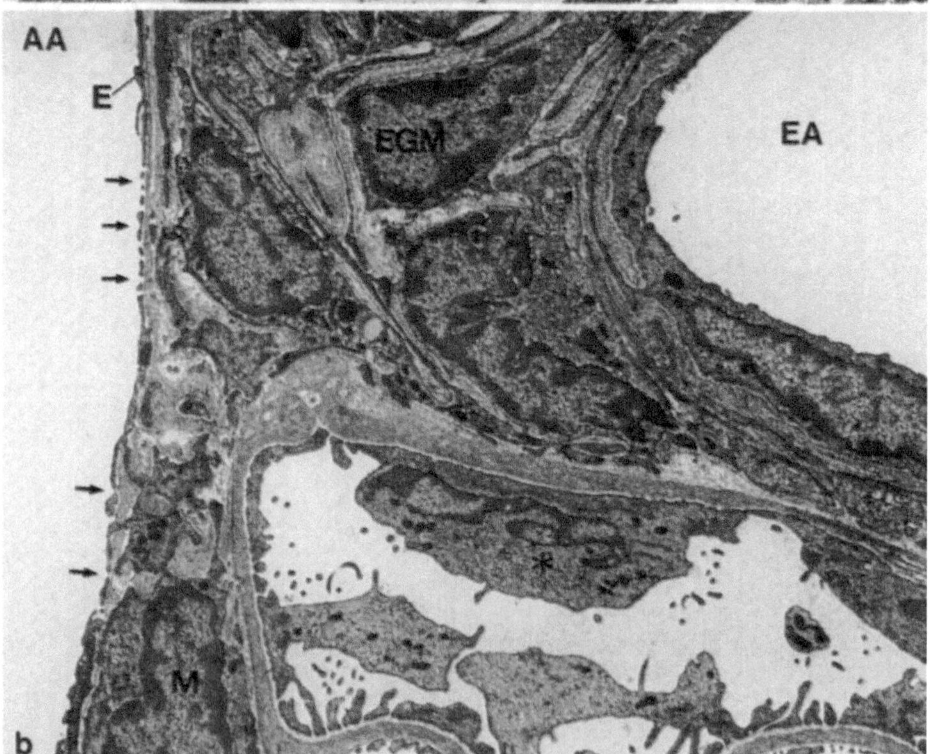
AA
E
EGM
EA
M
b

capillaries. They provide the typical filtration barrier established by a fenestrated endothelium (open pores), the GBM, and the interdigitating foot processes of the visceral epithelium (Figs. 13, 14). As is typical for the glomerular capillaries, the primary branches are also attached to the mesangium only on one, centrally located side.

3.2.2 Efferent Arteriole

The formation of the EA begins deep in the glomerular tuft and is completed when the efferent arteriole leaves the EGM. The EA may be subdivided into a beginning portion leaving the glomerulus – which we call the outflow segment, EA_{OS} – followed by a downstream or distal portion, EA_{DS}, in the cortical renal interstitium. The EA_{OS} may be further subdivided into an intraglomerular portion, EA_{OSi}, and a subsequent portion associated with the EGM, the EA_{OSe} (Fig. 16).

The EA_{OS} is established by the confluence of several tributaries (Fig. 10). More than one tributary may arise from one lobule. The confluence occurs successively. The majority of the tributaries unite deep in the tuft (distance to the opening of Bowman's capsule roughly 30–50 µm). By means of these initial confluences a trunk, the EA_{OSi}, is formed that clearly differs from glomerular capillaries because of its unique wall structure and location in the mesangium (Fig. 16). Later, further tributaries may join the EA_{OSi}. Quite common is a pattern with one additional tributary shortly before the EA_{OS} leaves the tuft (Fig. 17). The last tributary drains into the EA_{OS} at a level ranging between 5 and 30 µm off the exit from the glomerulus (the variability of the confluence pattern is discussed below). After its exit from the glomerulus, the extraglomerular portion of the outflow segment (EA_{OSe}) comes in broad contact with the EGM. Along the EA_{OSe} a smooth muscle layer is gradually formed.

Endothelium. The EA_{OSi} is formed by the confluence of several tributaries. The endothelium of the EA_{OS} is prominent by a high number of endothelial cells. Their cell bodies are preferably located directly at or shortly behind the confluence of tributaries to the EA_{OS} (Figs. 18–20). Behind the merging of the last tributary, endothelial cell bodies containing prominent nuclei often bulge into the lumen resulting in a narrow lumen. In favorable sections it can be seen that the endothelial cell bodies appear to be pushed downstream, possibly as the result of high shear stress at this site. In the distal portion of the EA (EA_{DS}) the endothelium regains its flat shape (Fig. 21).

The structure of the capillary wall already begins to change along the tributaries – sometimes up to 100 µm from the first confluence. The open endothelial pores characteristic of glomerular capillaries are progressively replaced by pores with diaphragms (Fig. 22a); finally, the pores disappear completely. A continuous

Fig. 15a,b. Afferent arteriole. **a** The peripheral wall portion facing the cortical interstitium contains smooth muscle and granular cells (*GC*). On the opposite side, smooth muscle cells are already replaced by extraglomerular mesangial (*EGM*) cells. The endothelium (*E*) of the AA towards the EGM is heavily fenestrated. The delineated area is seen at a higher magnification in **b**, showing the fenestrated endothelium. *M*, mesangium; *BC*, Bowman's capsule. TEM, **a** ×~900, **b** ×~6300

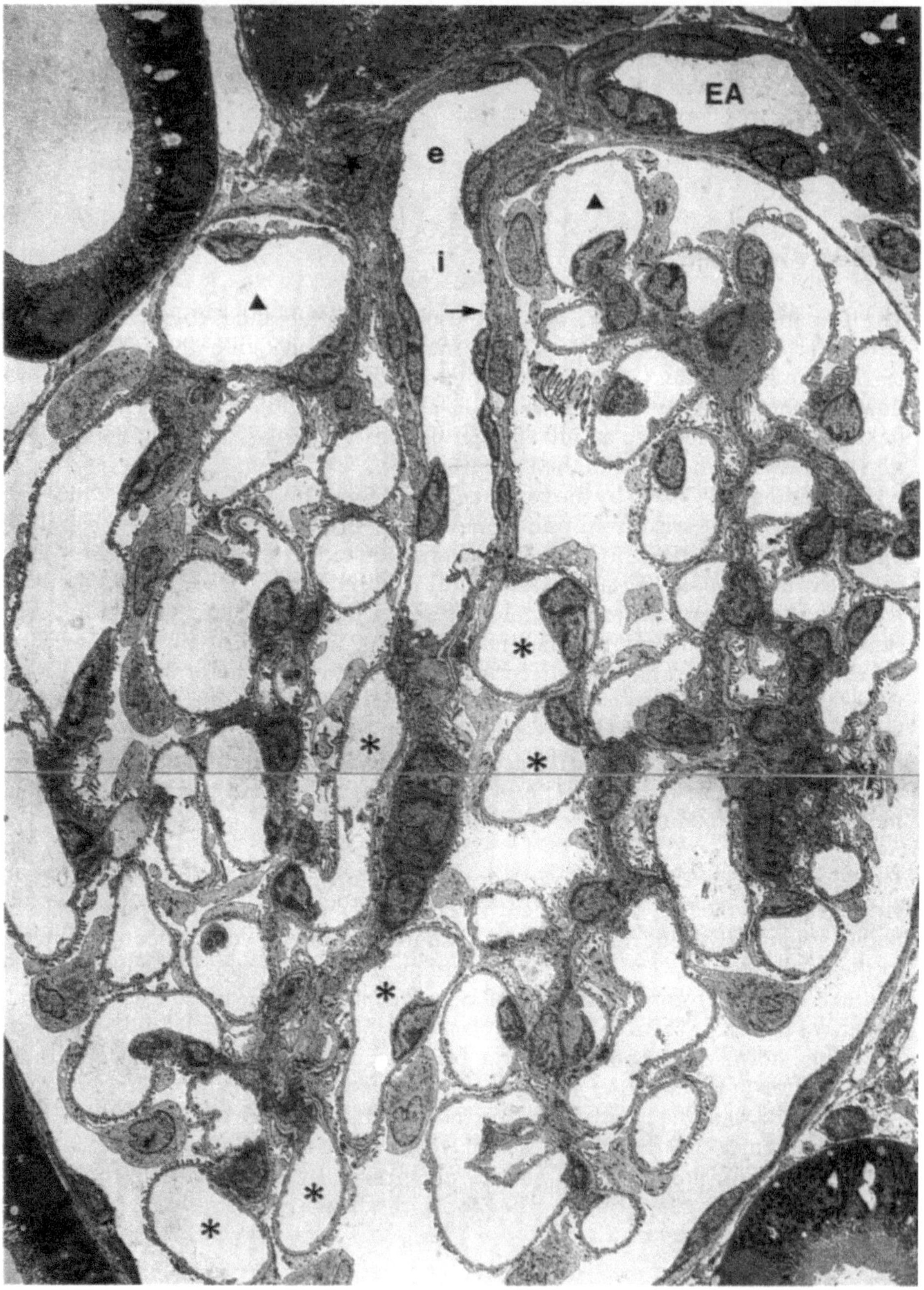
EA
e
i

endothelium with typical pinocytotic vesicles is present in the endportions of all tributaries (Fig. 22b,c) and is maintained throughout the EA_{OS} and the EA proper (Fig. 23). As long as the tributaries maintain a capillary wall structure with pores, they usually have wide lumina.

Structures Surrounding the Endothelial Tube. A small portion of the circumference of the glomerular capillaries has consistent contact with the mesangium. Beginning in the tributaries of the EA_{OS}, this contact region with the mesangium increases gradually by the penetration of mesangial tissue into the space between the endothelium and GBM (Fig. 18). Frequently, the endportions of the tributaries and, consistently, the EA_{OS}, are completely surrounded by a mesangial layer. This layer consists of the mesangial matrix and mesangial cell processes.

Along the EA_{OS}, single smooth muscle cells are encountered (Figs. 18, 19, 22d,e).[1]

The first smooth muscle cells are principally found at the confluence of the last tributary, and they are always located on the side facing Bowman's capsule epithelium. Towards the center of the glomerular stalk smooth muscle cells have not been encountered (Figs. 19, 10; see also 33). The smooth muscle cells associated with the EA_{OSi} were also identified by immunohistochemical demonstration of smooth muscle α-actin at this site (Elger et al. 1993).

At the exit from the glomerulus, the mesangium passes over into the extraglomerular mesangium. The EGM cells lying adjacent to the EA_{OSe} are elongated and form an incomplete layer around the vessel. EGM cell processes and intracellular microfilaments run in a circular direction around the EA_{OSe}. The smooth muscle cells progressively spread out as an additional layer between the endothelium and the EGM cells (Fig. 21). The muscle layer is complete, when the EA leaves the EGM (Fig. 23).

Along the EA_{OSi} the endothelium loses its close association with the glomerular basement membrane; the endothelial tube runs through the mesangium. An irregular layer consisting of basement-membranelike material (with an irregularly electron dense pattern), mesangial matrix, and occasional vesicles surrounds the endothelial tube (Fig. 22c). The basement-membranelike material protrudes into basal indentations of the endothelial cells. With this feature the EA_{OS} can usually be distinguished from the AA in sections of the vascular pole. Further downstream the EA (EA_{DS}) is surrounded by a smooth basement membrane.

[1] The smooth muscle cells can be distinguished morphologically from glomerular or extraglomerular mesangial cells by the following criteria (Fig. 22d,e): Smooth muscle cells run roughly perpendicularly to the long axis of the vessel, as do their microfilament bundles. Furthermore, smooth muscle cells are characterized by cytoplasmic densities associated with microfilaments, and by numerous invaginations of the plasma membrane (caveolae) oriented in rows. Microfilament bundles of mesangial cells are comparatively thinner. They are restricted to the periphery of the cell and are mainly located in cell processes extending in various directions. Caveolae are rarely encountered.

Fig. 16. Overview of a glomerulus showing the outflow segment of the efferent arteriole and its tributaries. The outflow segment comprises an intraglomerular (*i*) and extraglomerular portion (*e*); the latter passes the extraglomerular mesangium (*star*). Tributaries of the EA (*asterisks*) were identified by tracing them in the series of sections; they are located in a narrow medial sector of the tuft. The confluence of the last tributary (not hit in this section) occurs at the level indicated by an *open arrow*. Primary branches of the AA are marked by *triangles*. TEM, ×~1300

Fig. 17a-d. Longitudinal section series through an outflow segment and its tributaries. The series shows the intraglomerular portion of the outflow segment, EA_{OSI} (*1*), and a superficial tributary (*2*) joining the EA_{OSI} shortly before the exit (seen in the *lower right panel*). The EA_{OSI} is established by the confluence of capillaries *2a* and *2b* (*upper panels*); the superficial tributary (*2*) is formed by confluence of capillaries 2a and 2b (*upper left panel*). The EA_{OSI} (*1*), vessel *1a*, and vessel 2 have an endothelium with few proes (bridged by diaphragms) and a complete mesangial envelope. Vessels *2a* and *2b* are capillaries with open pores. Most capillaries belong to the efferent domain; the afferent capillaries (*black triangles*) have no connections to the efferent vessels. TEM, **a-d** ×~1350

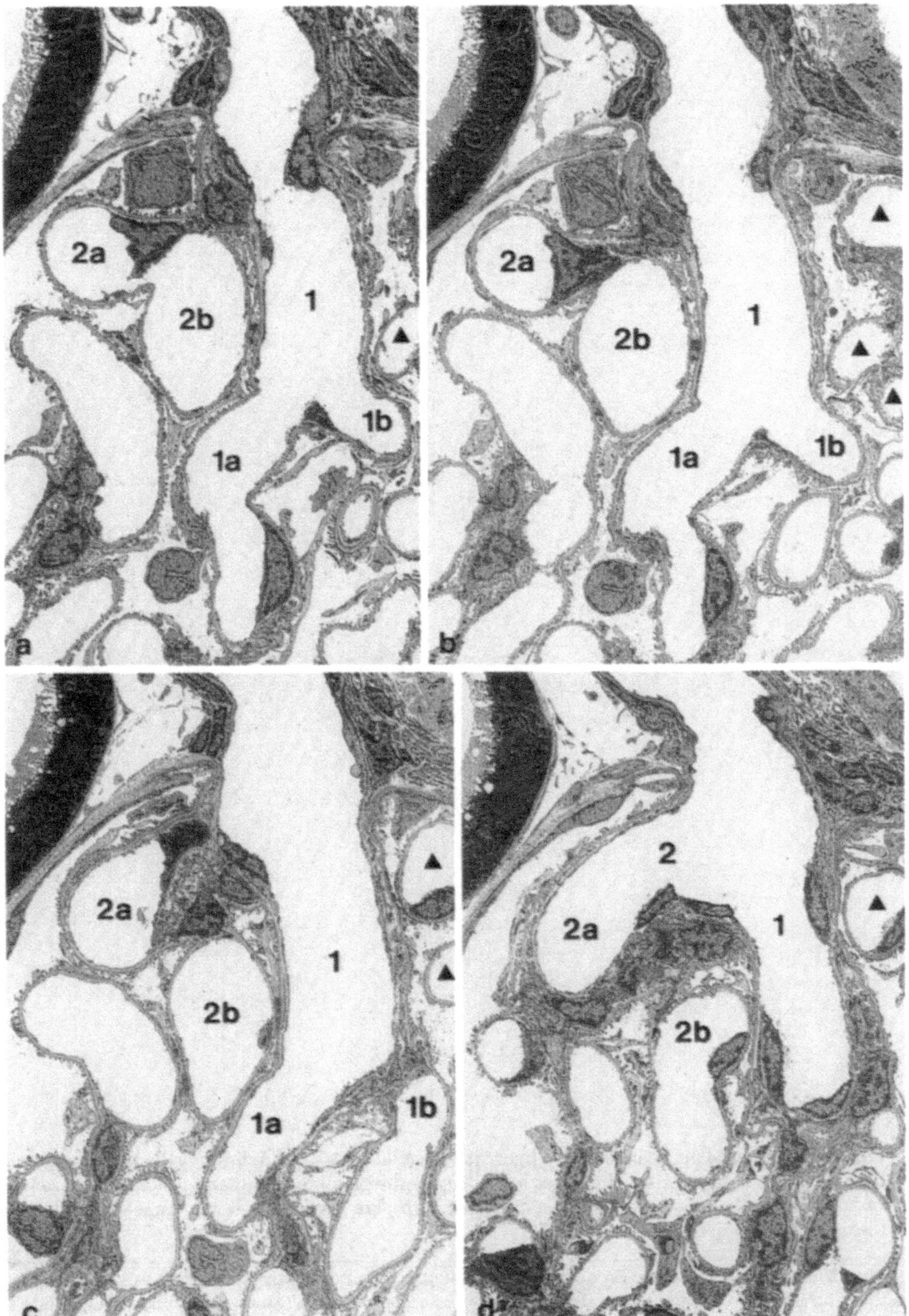
2a
2b
1
1a
1b
a
b
c
d
2

Figs. 18–21. This and the following TEM figures show serial cross sections through the outflow segment of the EA and its tributaries. The series begins inside the glomerular tuft and proceeds towards the vascular pole. Overviews (**a**) and detail pictures (**b**) are presented at the same magnification throughout the series

Fig. 18a,b. Tributaries to the EA approximately 40 μm inside the glomerulus. **a** The overview shows the central location of the two main tributaries (*asterisks*). The surrounding capillaries all belong to the afferent domain of the tuft. **b** The wall of these tributaries consists of an unfenestrated endothelium (*E*) with an irregular basal aspect and an incomplete mesangial envelope. A process of a smooth muscle cell (*SM*) is seen even at this deep intraglomerular level. *FP*, foot processes. TEM, **a** ×~1500, **b** ×~4200

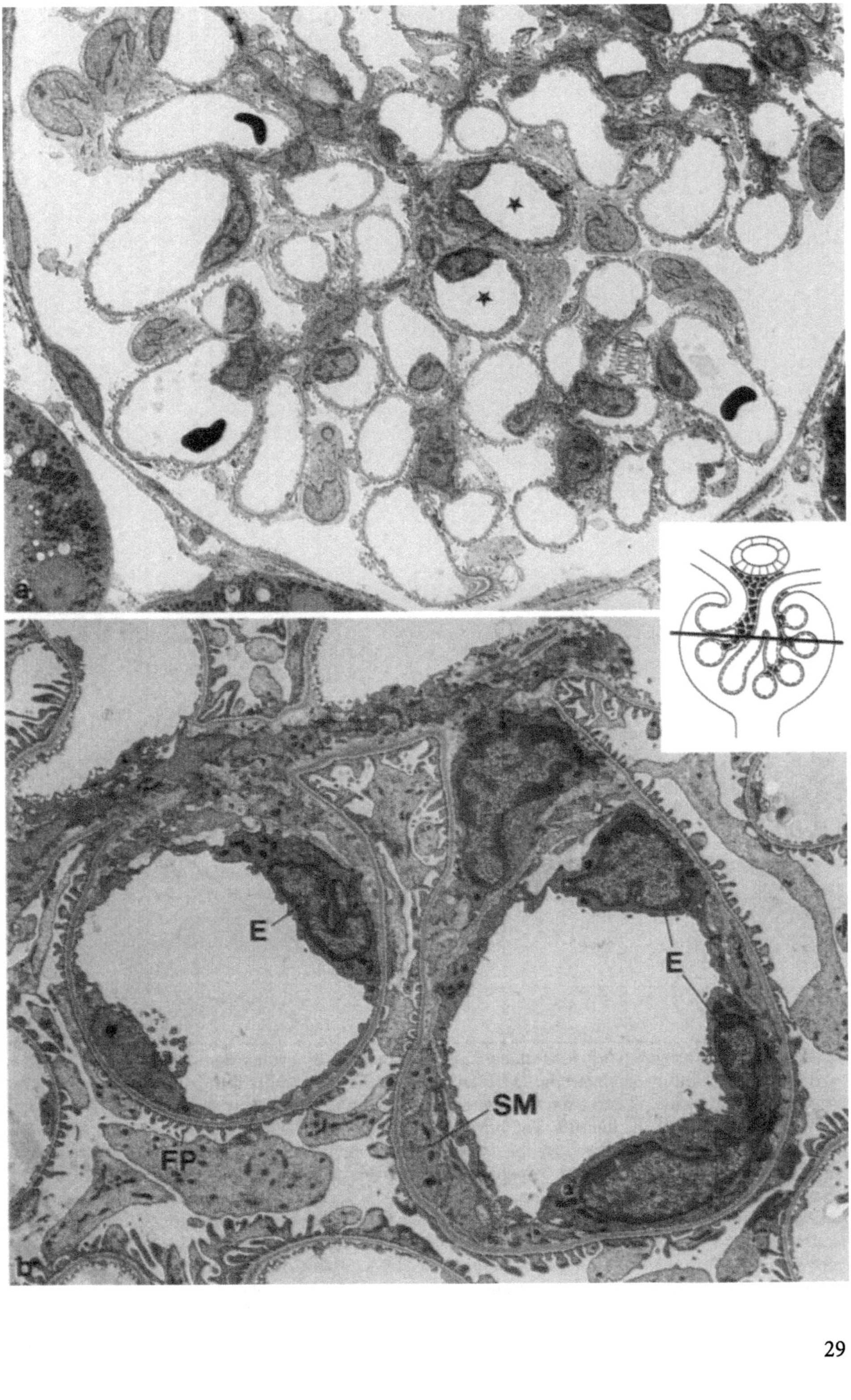
a
b
E
E
SM
FP

Fig. 19a,b. Outflow segment approximately 15 µm inside the glomerulus, demonstrating the "narrow portion" of the outflow segment.. **a** The afferent arteriole (*AA*) splits into primary branches. The branching point of the AA has a narrow spatial relationship to the outflow segment (*asterisk*), which is located in the center of the tuft. The outflow segment is enclosed – together with the AA – in a common compartment bordered by the GBM. **b** Higher magnification of the outflow segment in a subsequent section with several conspicuous features: the lumen is narrow; the continuous endothelium (*E*) consists of four cell bodies that bulge into the lumen; the endothelium is surrounded by a mesangial envelope made up of cells (*MC*) and matrix; a few smooth muscle cell processes (*SM*) are interspersed. AA and EA are separated only by mesangial tissue (*M*); there is no basement membrane separating the AA and EA. *GC*, granular cell; *P*, cell body of an outermost podocyte attached to the GBM by continuous cytoplasmic sheets. TEM, **a** ×~1500, **b** ×~4300

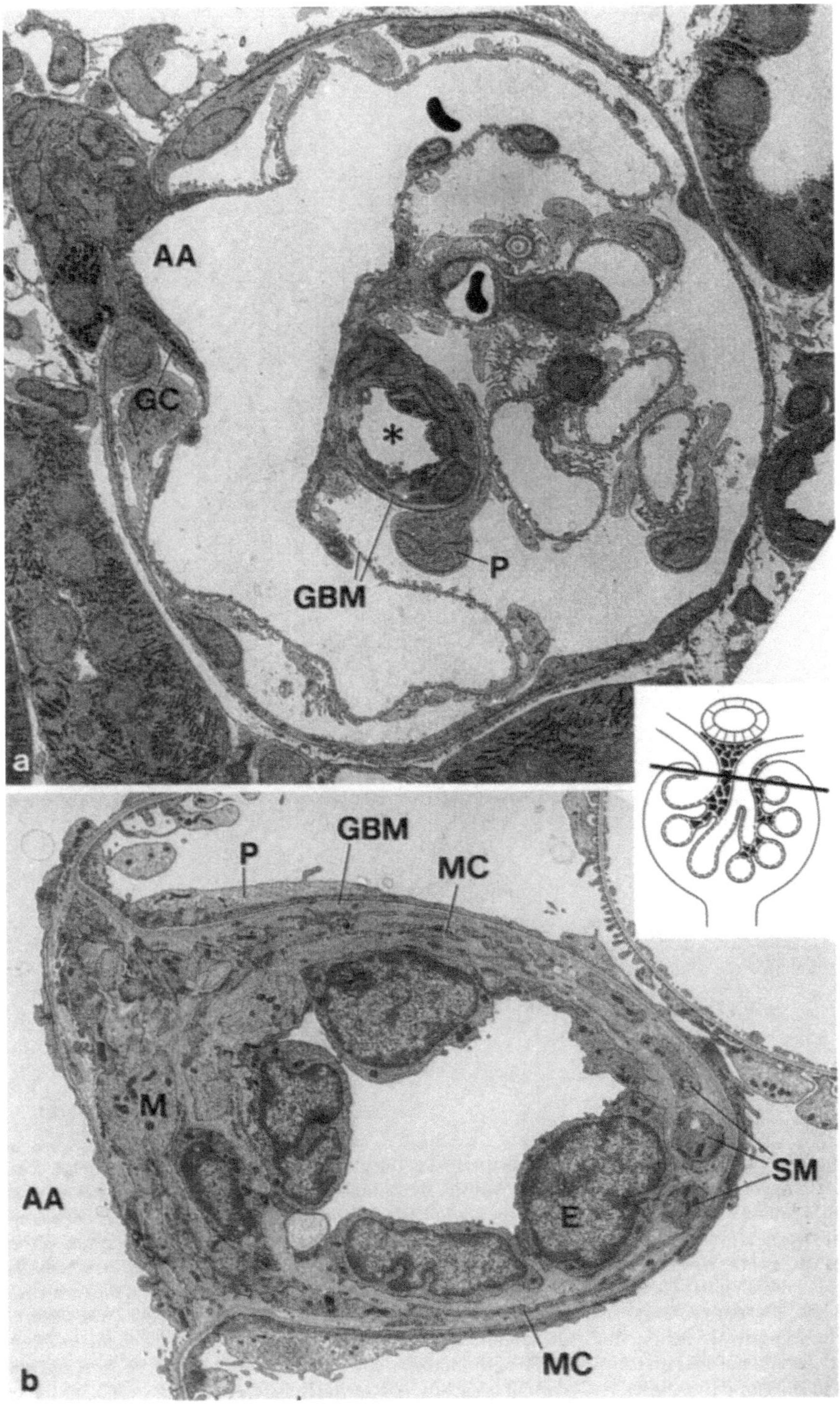
AA
GC
*
GBM
P
a
GBM
P
MC
M
AA
SM
E
MC
b

Fig. 20a,b. Outflow segment approximately 5 μm inside the glomerulus. **a** The overview shows a cross section of the outflow segment (*asterisk*) within the glomerular stalk. The cupolar portions of Bowman´s capsule are cut on both sides of the stalk. **b** The outflow segment displays an endothelium (*E*) with two prominent cell bodies, an incomplete smooth muscle layer (*SM*, surrounding about three-quarters of the circumference), and – outside the SM – a complete mesangial envelope consisting of cells (*MC*) and matrix. Towards the AA additional MC (*star*) are seen that represent the transition to the EGM. The entire structure is enclosed by the GBM, followed by epithelial cells (transition of podocytes to parietal cells). A thick microfilament bundle (*arrows*) is seen inside such an epithelial cell. In the epithelial cells on the other side of the stalk no such bundle is seen in this section, but was present in subsequent sections. *BC*, Bowman's capsule; *US*, urinary space. TEM, **a** ×~1500, **b** ×~4300

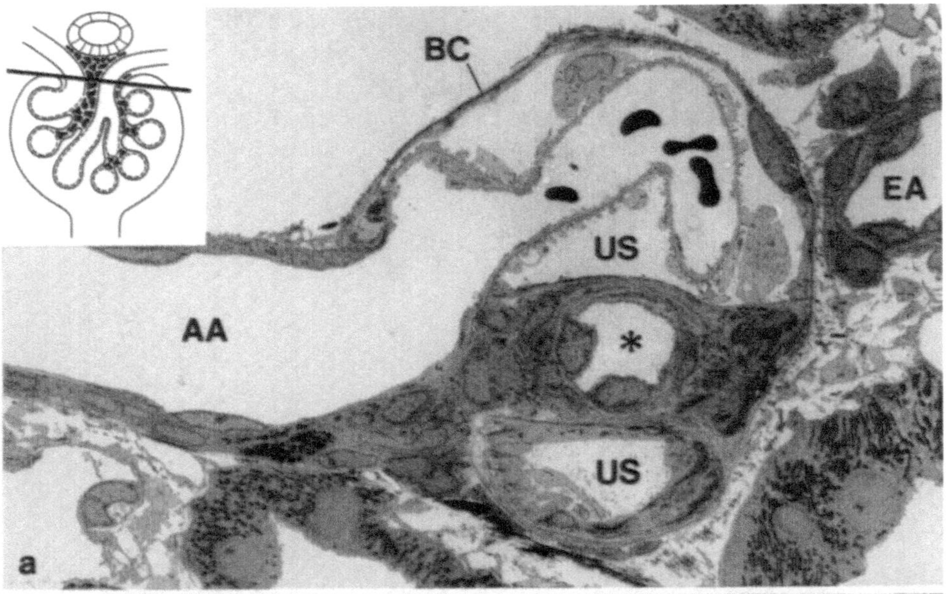
BC
EA
US
AA
*
US
a

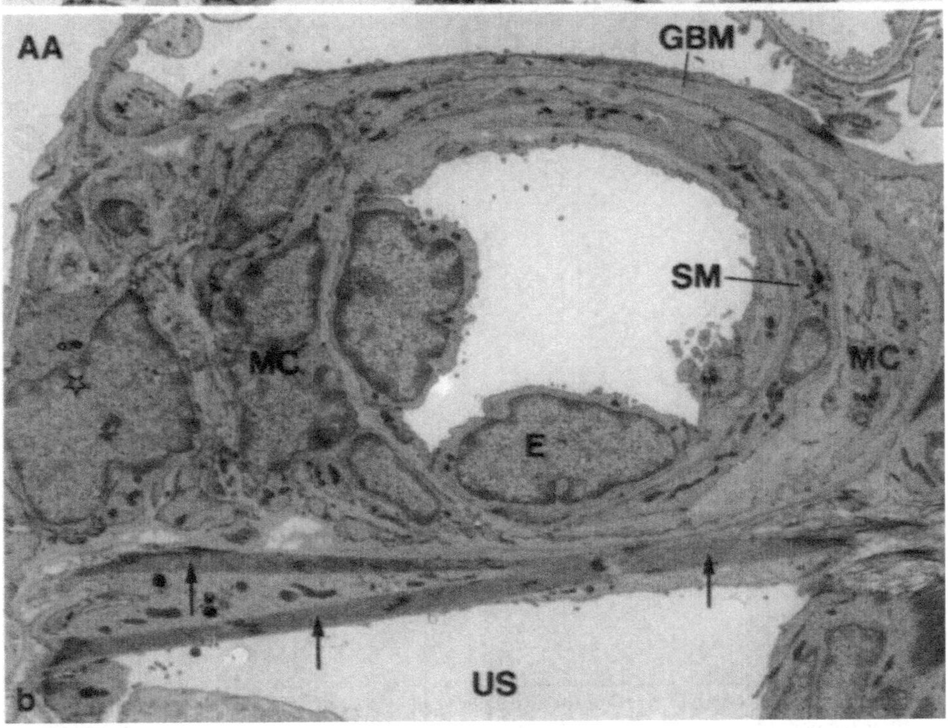
AA
GBM
SM
MC
MC
E
US
b

Fig. 21a,b. Outflow segment passing the extraglomerular mesangium (EGM). **a** In addition to the outflow segment (*asterisk*), the EA and AA are seen outside the glomerulus. AA and outflow segment are separated by EGM cells. **b** The outflow segment consists of a continuous endothelium (*E*), followed first by an incomplete smooth muscle layer (*SM*) and, second, by a complete envelope of EGM cells. In contrast to the situation in Fig. 19, the endothelium is flat; it is closely associated with the surrounding basement membrane material by protrusions and indentations (*small arrow*). *GC*, granular cell; *US*, urinary space. TEM, **a** ×~1500, **b** ×~4300

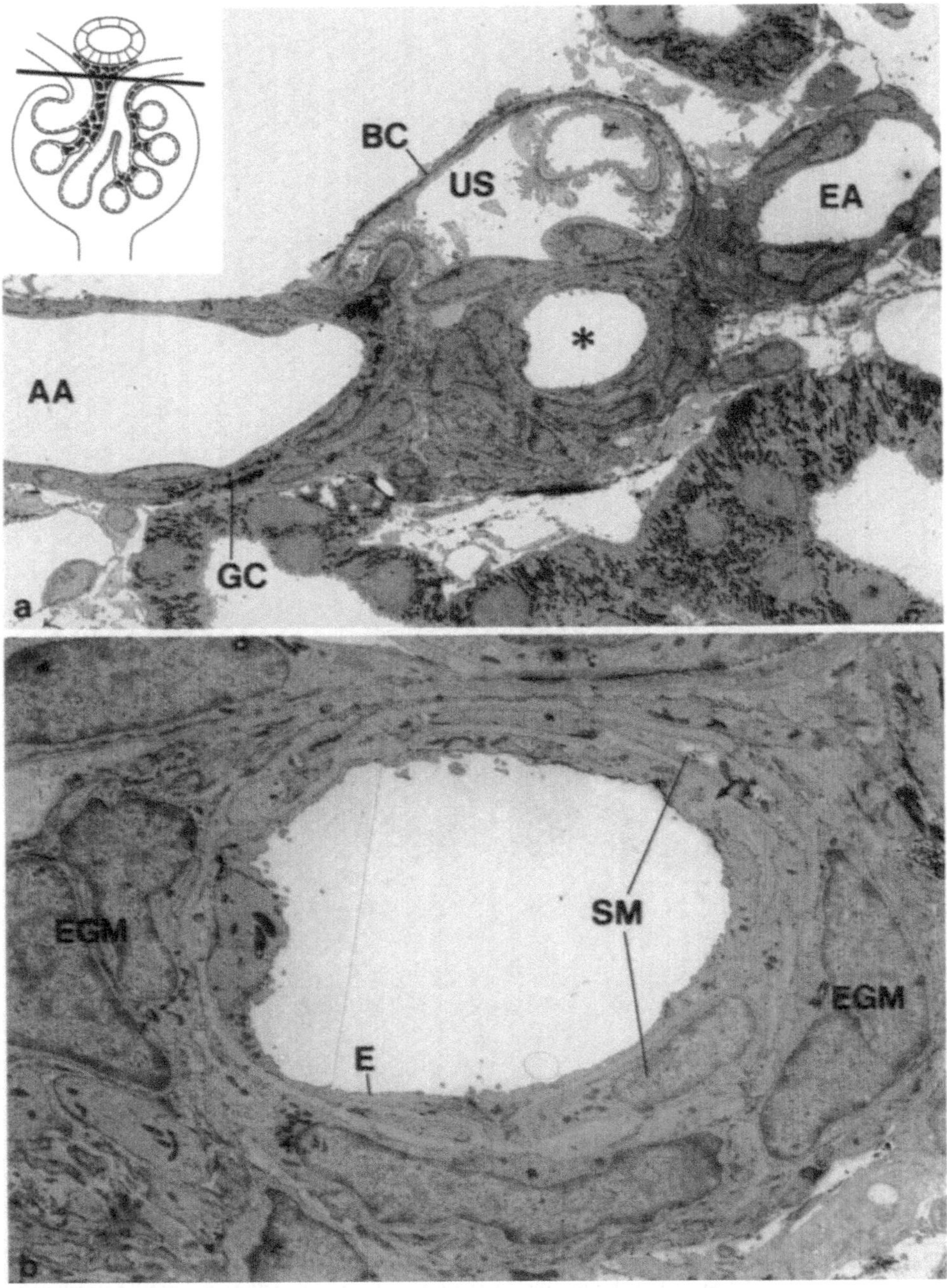
BC
US
EA
*
AA
GC
a
EGM
SM
EGM
E
b

Fig. 22a–c. Details of wall structure of the intraglomerular outflow segment and its tributaries. **a** Wall of a tributary located deep in the glomerulus. In contrast to the usual picture of a capillary wall with open pores (*arrows*), the endothelial pores of tributaries (at the *top of the figure*) are often bridged by diaphragms (*arrowheads*). *C*, capillary lumen; *PO*, podocyte layers; *US*, urinary space. **b, c** Wall of tributaries close to the junction with the trunk segment, as seen in Fig. 18b. The wall consists of a continuous endothelium (*E*) with numerous pinocytotic vesicles (*arrows*). Later, a mesangial layer (mainly matrix, *M*) penetrates between endothelium and GBM (shown in **c**). Note irregular basal aspect of the endothelium typical of the outflow segment. TEM, **a, b.** ×~34 200, **c** ×~36 100.

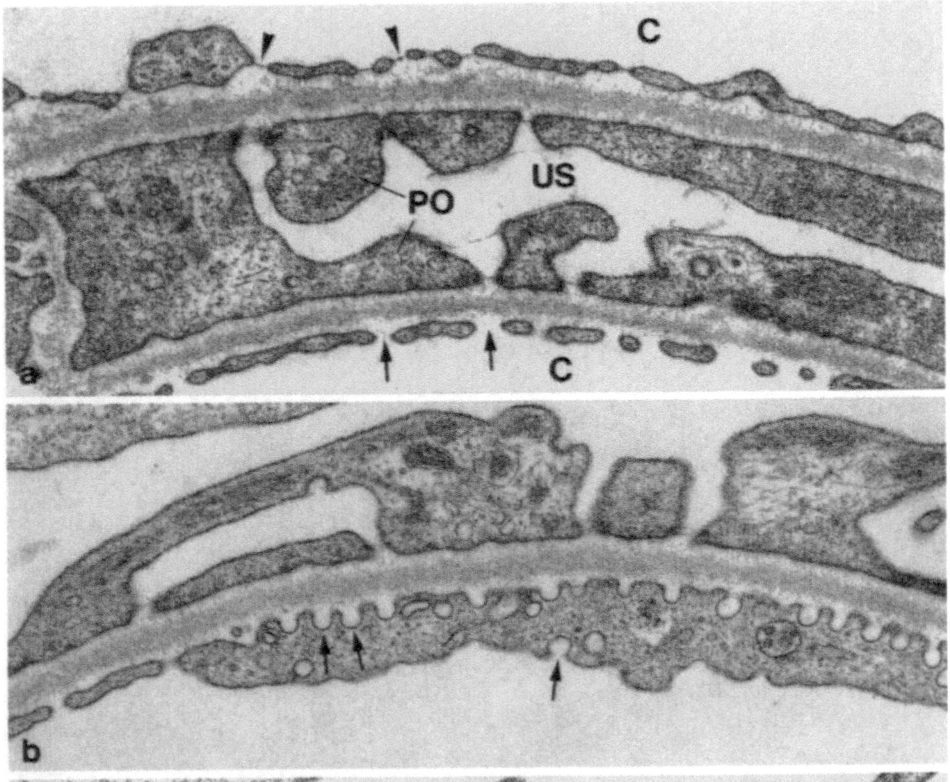
C
PO
US
a
C
b
GBM
M
M
c

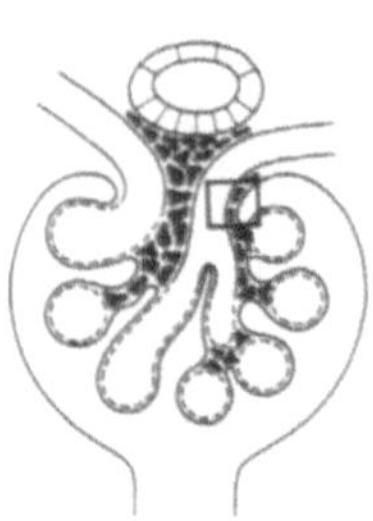

Fig. 22 d, e. Wall of the outflow segment distal to the last confluence. The wall consists of a continuous endothelium (*E*), a layer of smooth muscle cells (*SM*), and mesangial cells (*MC*) and mesangial matrix, followed by the GBM and podocytes. Two myoendothelial junctions are seen (*small arrows*). Note outermost podocyte (*asterisk*), which is directly attached to the GBM without foot processes. The basal cytoplasm contains conspicuous bundles of microfilaments (*large arrow*). *US*, urinary space. **e** Wall of outflow segment (level as seen in Fig. 20b) containing smooth muscle cells (*SM*) mesangial cells (*MC*). The SM are arranged circularly around the vessel lumen with circularly running microfilament bundles (*F*). In contrast, in MC microfilament bundles are mainly found in cell processes running in all directions. The SM exhibit rows of caveolae along the cell membrane (*small arrows*). *E*, endothelium of outflow segment. TEM, **d** ×~11 400, **e** ×~18 200,

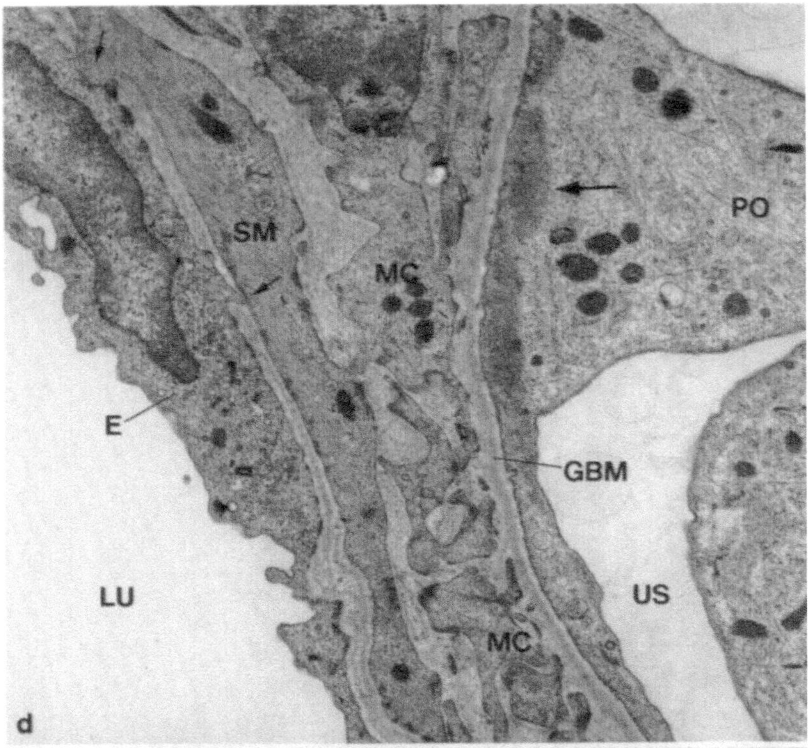
SM
MC
PO
E
GBM
LU
US
MC
d

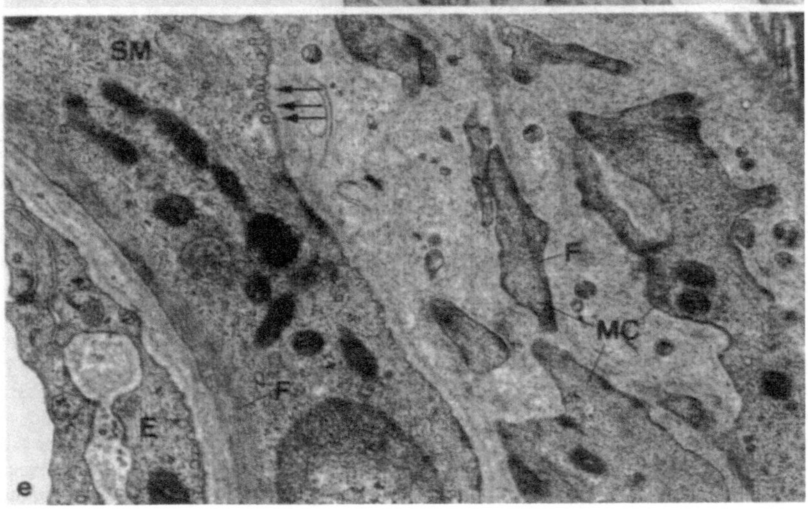
SM
F
MC
F
E
e

Fig. 23. Schematic drawing of a longitudinal section showing the structural organisation of afferent and efferent arterioles (*AA*, *EA*) at the vascular pole. Immediately after its entry into the glomerulus, the AA divides into several primary branches, which exhibit the characteristic wall structure of glomerular capillaries, including a fenestrated endothelium. Even the endportion of the AA where it faces the extraglomerular mesangium (EGM) often has a fenestrated endothelium (*small arrows*). The outflow segment of the EA (EA_{OS}) is formed by confluence of several tributaries inside the glomerular tuft. It has an intraglomerular and an extraglomerular portion (EA_{OSi}, EA_{OSe}). The outflow segment is characterized by a continuous endothelium with densely packed cells. Inside the glomerulus, it has a complete mesangial envelope and close spatial relationships to the AA. The morphological changes along the tributaries and the portions of the outflow segment are summarized as follows:
Tributaries (*T*) to the outflow segment: most tributaries are typical glomerular capillaries with open endothelial pores and a restricted interface to the mesangium. In the endportion of tributaries a complete mesangial envelope is established. Diaphragms appear in the endothelial pores at variable distances from the confluence.
Intraglomerular portion of the outflow segment (EA_{OSi}): The intraglomerular portion may be subdivided into a part before and after the last confluence. A complete mesangial envelope is present throughout. The endothelium is rich in cells and has lost its porous character. After the last confluence smooth muscle cells appear.
Extraglomerular portion of the outflow segment (EA_{OSe}) in the EGM: This segment is associated with the EGM; the smooth muscle layer is still incomplete.
Distal segment of the EA in the cortical interstitium (EA_{DS}): A complete smooth muscle layer is established. The endothelium is smooth. The glomerular basement membrane (*GBM*), together with the transitional part of the epithelium, borders the glomerular stalk, which contains the endportion of the AA as well as the EA_{OSi}. Both are separated only by a narrow layer of mesangial tissue. *BC*, Bowman's capsule; *MC*, mesangial cell; *MM*, mesangial matrix

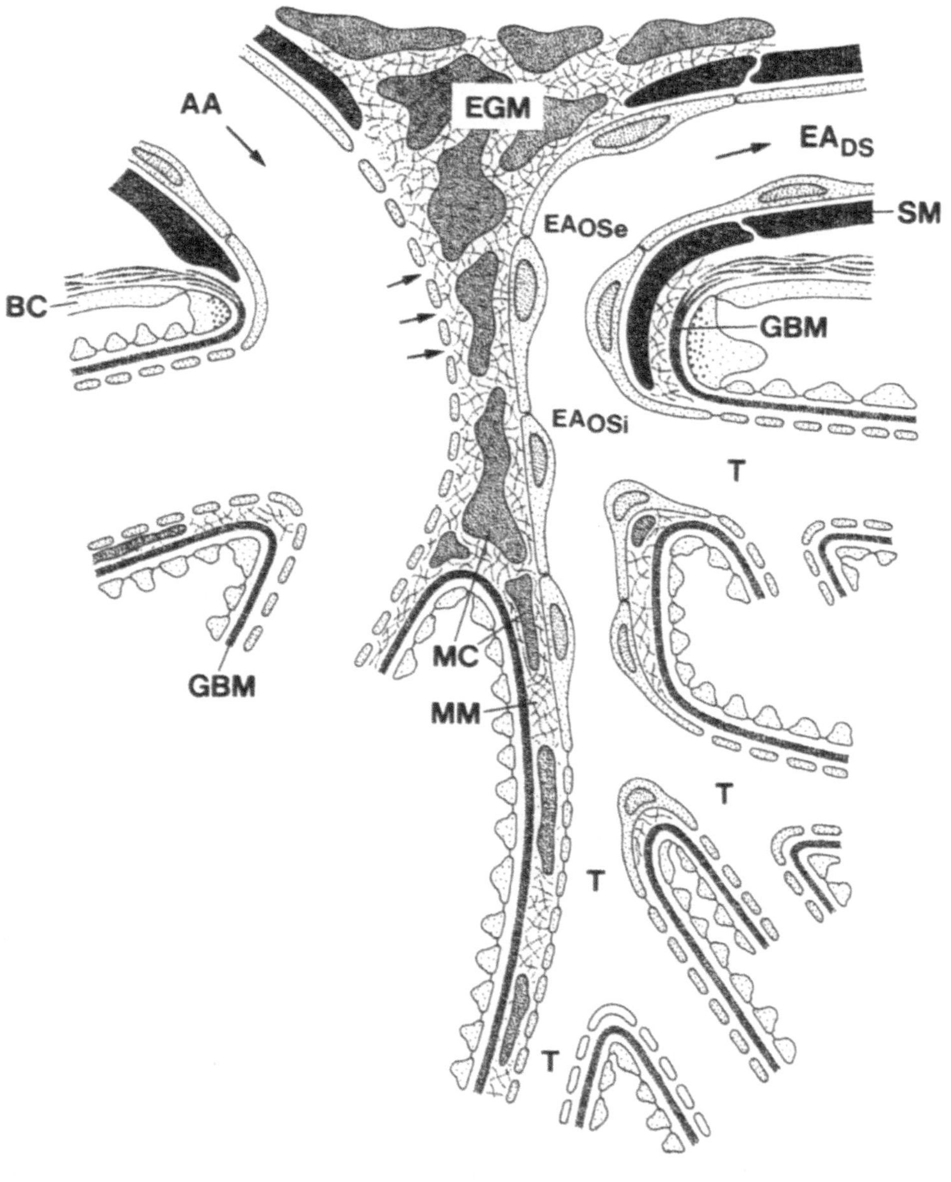
EGM
AA
EA$_{DS}$
SM
EA$_{OSe}$
BC
GBM
EA$_{OSi}$
T
GBM
MC
MM
T
T
T

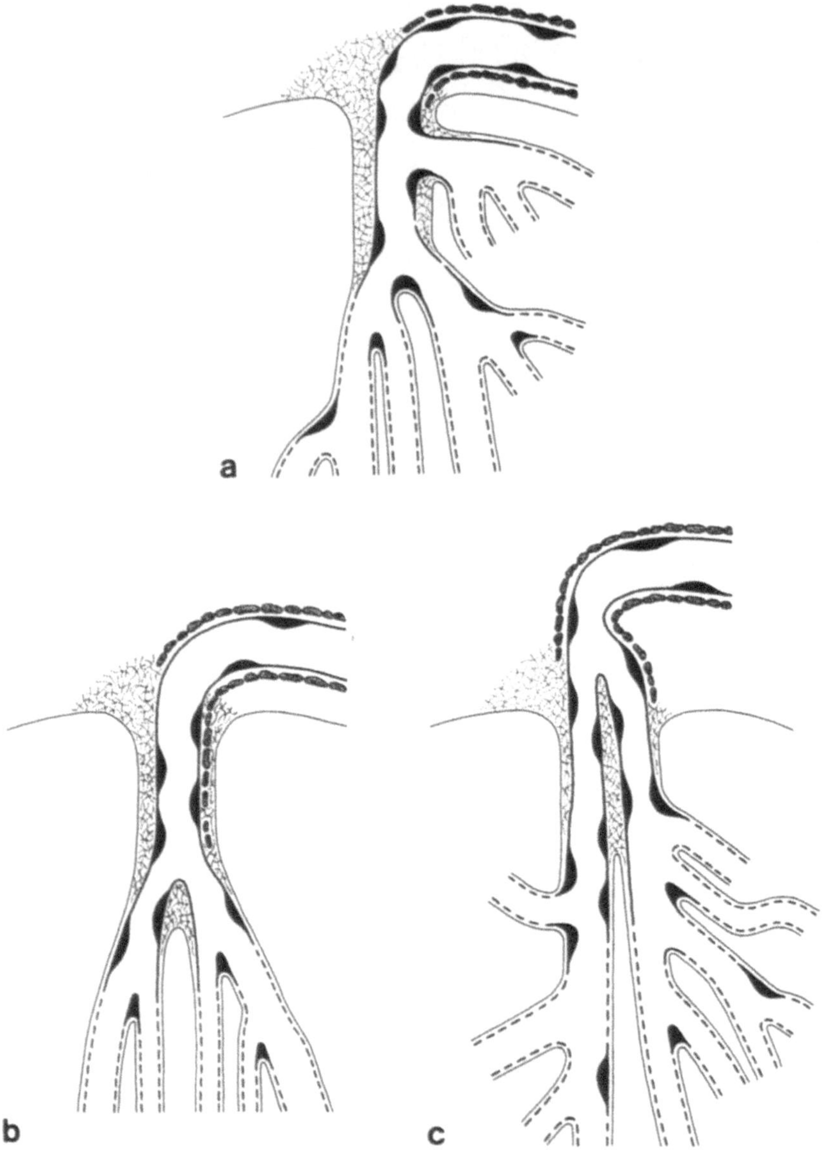

Fig. 24 a–c. Schematic drawing showing the structural variability of the outflow segment. The tributaries draining the capillary network merge successively. Thirty to forty micrometers off the vascular pole, the trunk (sometimes two) of the EA is established. It has a continuous endothelium surrounded by a complete layer of mesangium. The trunk may receive one or, rarely, two further tributaries. The level where the last tributary joins is variable. In most cases, this confluence is situated within 5—10 μm off the glomerular exit (shown in **a**), but it may also be located 30 μm before the exit (**b**). Occasionally, the last fusion occurs outside in the extraglomerular mesangium (**c**). Rarely, two completely separate efferent arterioles are formed (not shown). Smooth muscle cells are never encountered in segments before the last confluence

Variability in the Origin of the Efferent Arteriole. The EA_{OS} shows some variability with respect to the confluence pattern of the tributaries (Fig. 24). Principally, the confluence occurs successively. However, the extension of the confluence region is variable: In a minor proportion of glomeruli, the confluence of all tributaries is already accomplished about 25 μm off the vascular pole, so that the entire confluence occurs almost at one point deep in the tuft (Figs. 19, 24b). In the majority of the glomeruli, however, the confluence extends over a greater distance: In addition to the tributaries which have merged deep in the tuft, generally one, rarely two tributaries join the EA_{OSi} much closer to the exit (Figs. 16, 17, 24a). Occasionally, the fusion of the last tributary is located beyond the exit (Fig. 24c); in rare cases, two completely separate efferent arterioles are found.

The location of the final fusion determines several differences in the wall structure of the EA_{OSi}. First, as smooth muscle cells were never found before the final confluence (which is occasionally located at the exit of the glomerulus), smooth muscle cells along the EA_{OSi} are present in many but not all glomeruli (compare Figs. 17 and 19). Second, as described above, an especially high number of endothelial cells are found bulging into the lumen behind the confluence of the last tributary. Thus, dependent on the varying location of the final confluence, this narrow portion also has a variable location: it is situated either inside the glomerulus, at the exit, or outside (Fig. 24).

The variability of the confluence pattern to the EA_{OS} is in part dependent on the number and arrangement of the glomerular lobules. Three lobules have been identified in most glomeruli from rats, but also two, four or five lobules can occur (Murakami 1972; Kikuta and Murakami 1989; Winkler et al. 1991). The lobules are drained usually by one tributary (sometimes by two or more) to the EA_{OS}, but also by vessels draining into the capillary network of other lobules. Moreover, the lobules are not equal in size.

Dimensions of the Efferent Arteriole. As described above, the length of the entire intraglomerular portion of the outflow segment (EA_{OSi}) is usually about 50 μm. The level of the last confluence to the EA_{OSi} ranges between a few and 30 μm off the exit. The portion associated with the EGM, EA_{OSe}, is approximately 20 μm long.

The width of the EA_{OS} and the tributaries was estimated in several superficial glomeruli series of semithin and ultrathin sections. The lumen were measured in these vessels, which were cut in a plane perpendicular to the long axis of the vessels. The width of the efferent arteriole was always smaller than that of the corresponding AA, as has been previously shown by others (Dieterich 1978; Unehira 1981; Kriz and Bachmann 1985; Hitomi et al. 1987). A conspicuously narrow portion was generally found just downstream from the last point of confluence. This portion was distinctly narrower than the subsequent portion of the EA in the renal interstitium (EA_{DS}). The narrow portion had an axial length of approximately 10 μm. The narrowness appears in part to be derived from the high number of endothelial cells (see above, Fig. 19). The width of the narrow portion may be as small as 4.5 μm, whereas the subsequent portion in the EGM (EA_{OSe}) may have a diameter of approximately 7.5 μm (Fig. 21).

3.3
Extraglomerular Mesangium

The mesangium passes through the opening of Bowman's capsule and continues into the extraglomerular mesangium (EGM; Goormaghtigh cells, lacis cells). Some 20 cells and extracellular matrix make up the EGM; generally no further structures are included in this field.

The EGM is a roughly cone-shaped structure. It lies above the opening of Bowman's capsule between the two arterioles and extends onto the parietal basement membrane (Fig. 25, 26). In this way, the EGM is attached to the PBM as well as to the walls of both arterioles. On the opposite side, the macula densa of the thick ascending limb is broadly

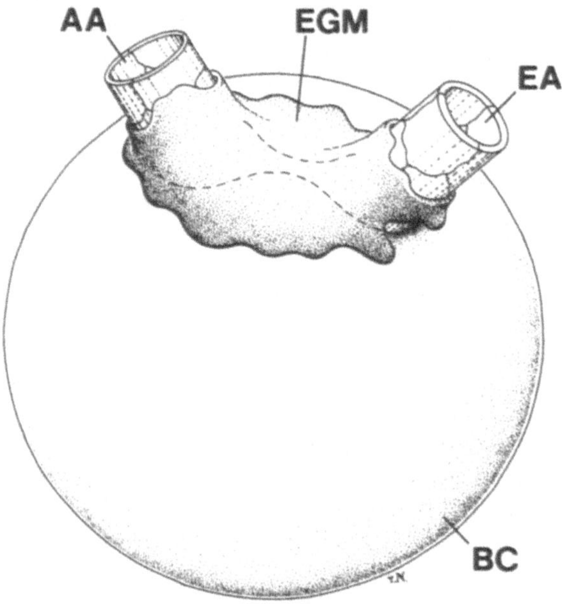

Fig. 25. Schematic of the extraglomerular mesangium (*EGM*). The glomerulus is shown as a globe. Its outer aspect is represented by the parietal basement membrane of Bowman's capsule (*PBM*). The EGM lies between the two arterioles above the opening of Bowman's capsule (*broken line*). It is attached to the PBM and has extensive contacts with the two arterioles. The macula densa and the smooth muscle layers of the arterioles are not shown

Fig. 26. Interarteriolar section through the vascular pole, displaying the glomerular stalk bordered by the cupolas of Bowman's capsule (*BC*). The EGM cells are arranged in layers parallel to the base of the macula densa (*MD*) to which they are broadly attached. They fill the funnel of the GBM reflection, apparently interconnecting opposing parts of the funnel wall. When approaching the BC, the EGM cells fall apart into processes, which attach to the parietal basement membrane (*PBM*). The transition from the GBM to the PBM is covered by the outermost podocyte (*asterisk*). Note that interdigitating foot processes extend far beneath the EGM. TEM,×~3750

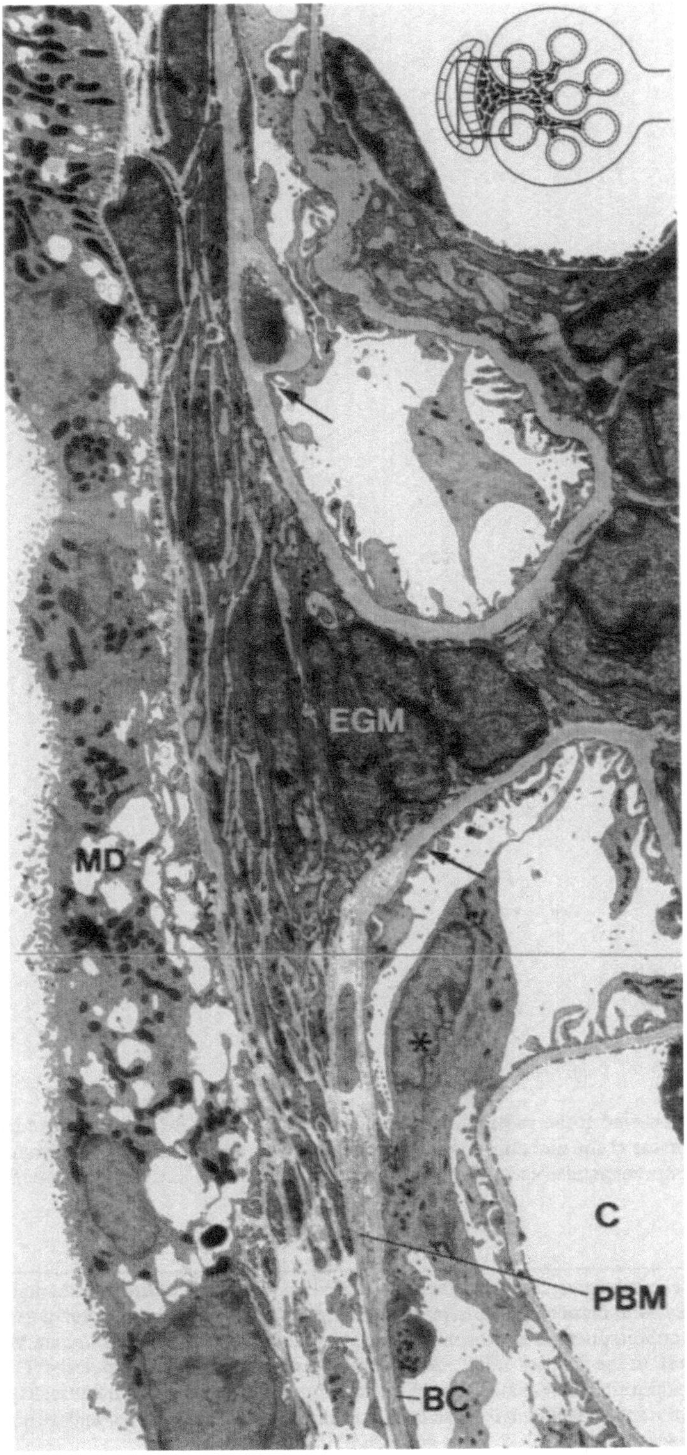
EGM
MD
C
PBM
BC

Figs. 27–30. Cross-sectional series of the vascular pole in a plane roughly parallel to the base of the macula densa.The series begins at the macula densa and proceeds towards the opening of Bowman's capsule. The afferent and efferent arterioles (*AA*, *EA*) are always seen at the *top* and the *bottom* of the figures, respectively

Fig. 27. Superficial part of the extraglomerular mesangium (*EGM*) and the base of the macula densa (*MD*).The EGM consists of a solid mass of cells and matrix, extending between both arterioles. On the *left side* the section hits a cupola portion of Bowman's capsule (*BC*) and, on the *right side*, the MD (delineated by a *broken line*). In the vicinity of BC, the EGM cells form numerous cell processes (*CP*). The boundary to the cortical interstitium (*CI*) is seen in the *upper and lower parts* of the figure; EGM cell bodies (*asterisks*) face the interstitium. The EGM cells are mainly roundish or polygonal with no apparent orientation in this section plane. *US*, urinary space. TEM, ×~3500

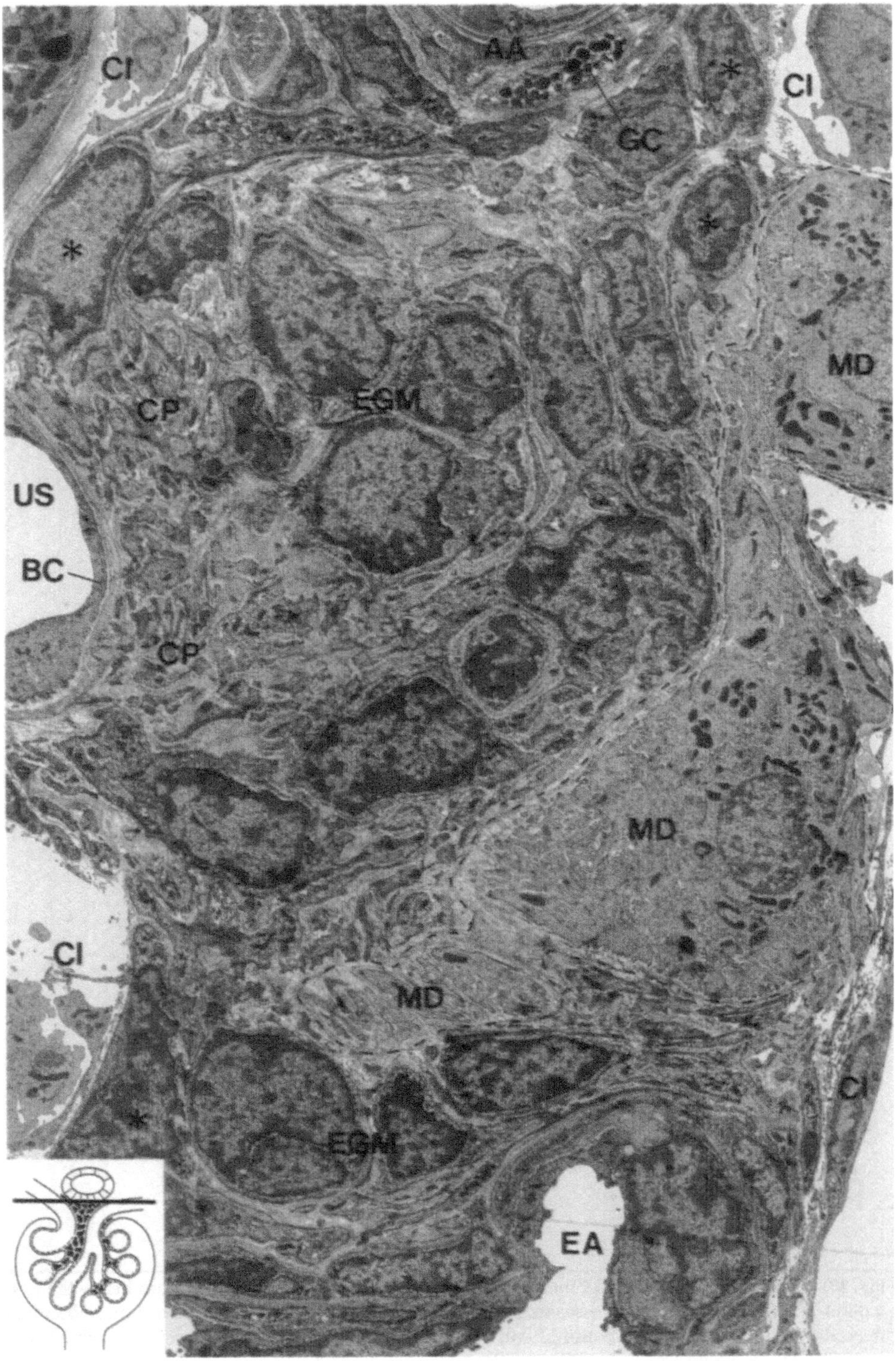

CI
AA
GC
*
CI
*
*
MD
CP
EGM
US
BC
CP
MD
CI
MD
*
EGM
CI
EA

Fig. 28. Superficial part of the EGM beneath the macula densa (**MD**), roughly 0.5 μm below the first section in Fig. 27. EGM cell processes extend in all directions, forming an irregular pattern between the cell bodies. Note that the AA and EA still have a complete smooth muscle layer (*SM*) adjacent to the EGM. *BC*, Bowman`s capsule; *GC*, granular cell. TEM, ×~3600

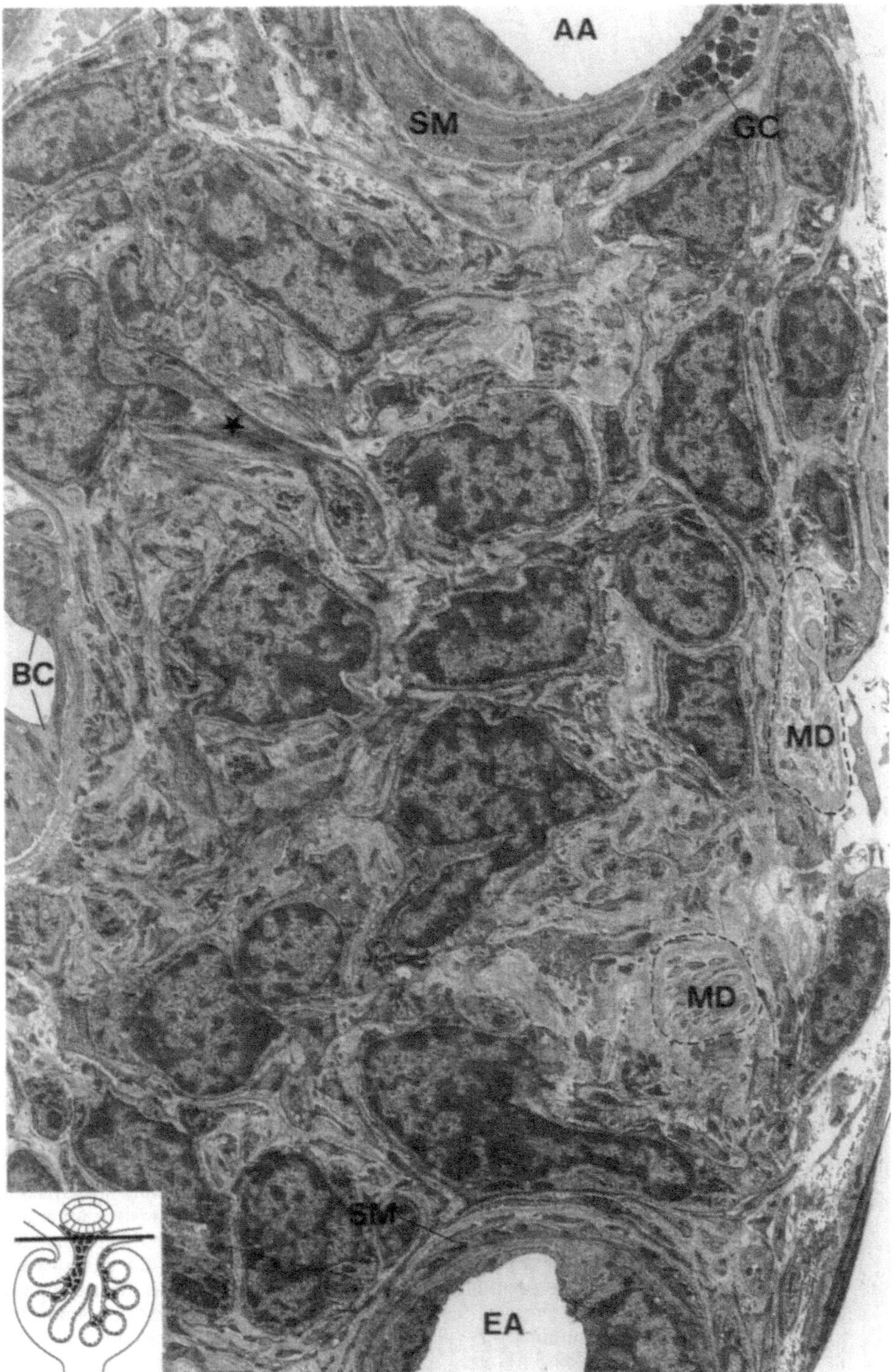
AA
SM
GC
BC
MD
MD
SM
EA

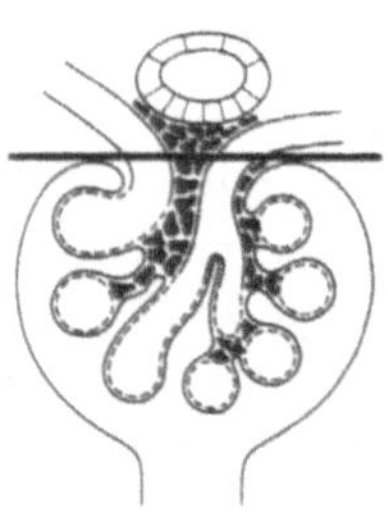

Fig 29. Deep portion of the EGM. AA and EA and the cupola of Bowman's capsule (*BC*) on one side are cut. The middle portion containing the large cell bodies represents that part of the EGM that will transform into the glomerular stalk. The right part of the EGM, consisting almost completely of cell processes, is situated just above the right cupola portion of BC to which the cell processes anchor (the corresponding part of BC seen in the next figure is delineated by a *broken line*). At the AA, the muscle layer (*SM*) becomes discontinuous; the basement membrane merges with the EGM matrix (*arrow*) (see also Fig. 32). At the EA, an EGM cell (*asterisk*) extends around the vessel outside the SM layer. TEM, ×~3450

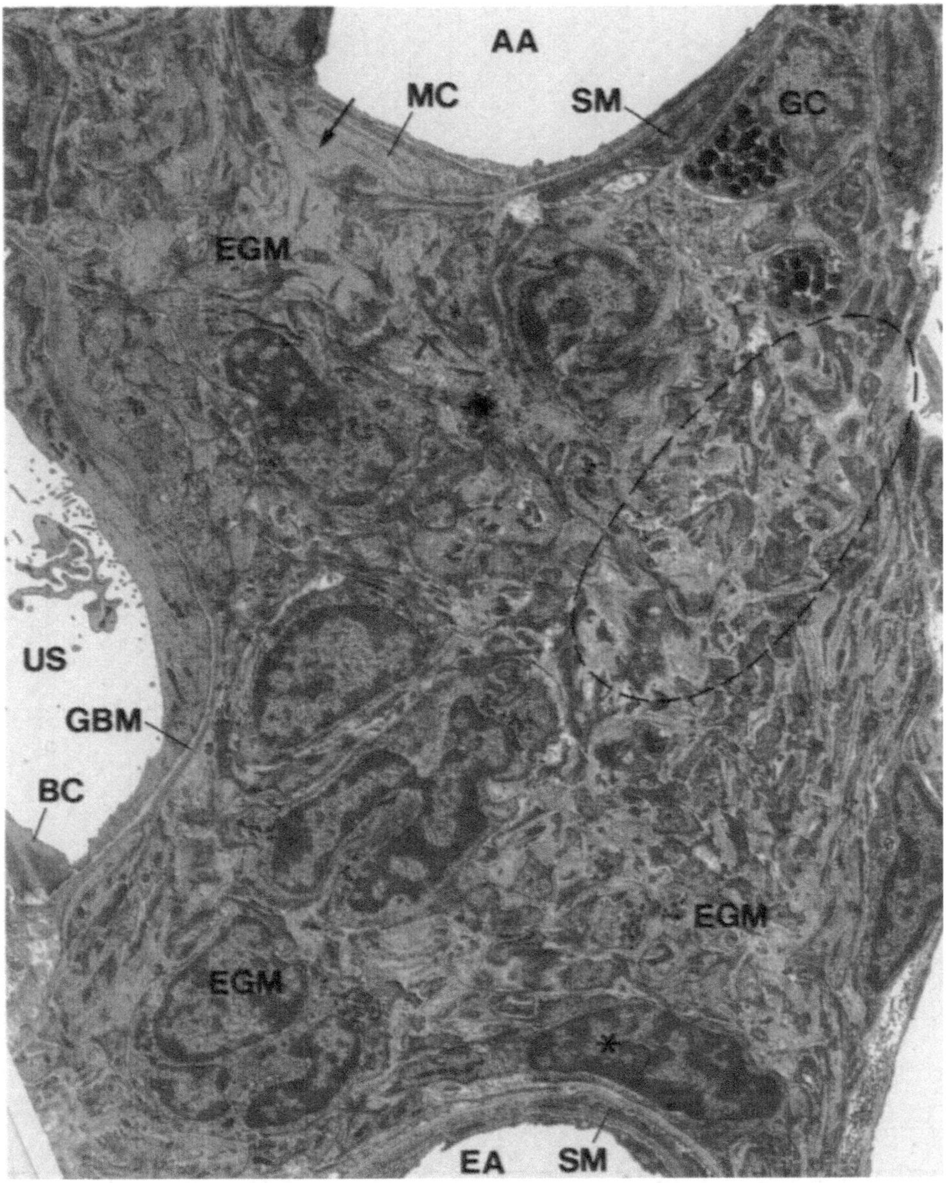
AA
MC
SM
GC
EGM
US
GBM
BC
EGM
EGM
EA
SM

Fig. 30a,b. Deep portion of the EGM with its attachment to the parietal basement membrane (*PBM*) and to the transition into the GBM. **a** Higher magnification of a section subsequent to Fig. 29 showing the right cupola of Bowman's capsule. Part of a parietal cell (*PE*) is already cut. Note two different aspects of the basement membrane (*GBM* and *PBM*). Both GBM and PBM are bordered by EGM cell processes (*CP*), which are connected to the basement membrane. In most EGM cell bodies and in primary processes dense accumulations of microfilaments are seen beneath the plasma membrane (*arrows*). At this level the EA is surrounded by an EGM cell (*asterisk*) and its processes (*double arrows*). Smooth muscle cells are lacking at this site. **b** Contact between EGM and GBM. A primary process (*PP*) divides into several small secondary processes (*SP*), which contain thick microfilament bundles and are anchored to the GBM (*arrows*). *EM*, extraglomerular mesangial matrix; *N*, nucleus of EGM cell. TEM, **a** ×~4750, **b** ×~27 000

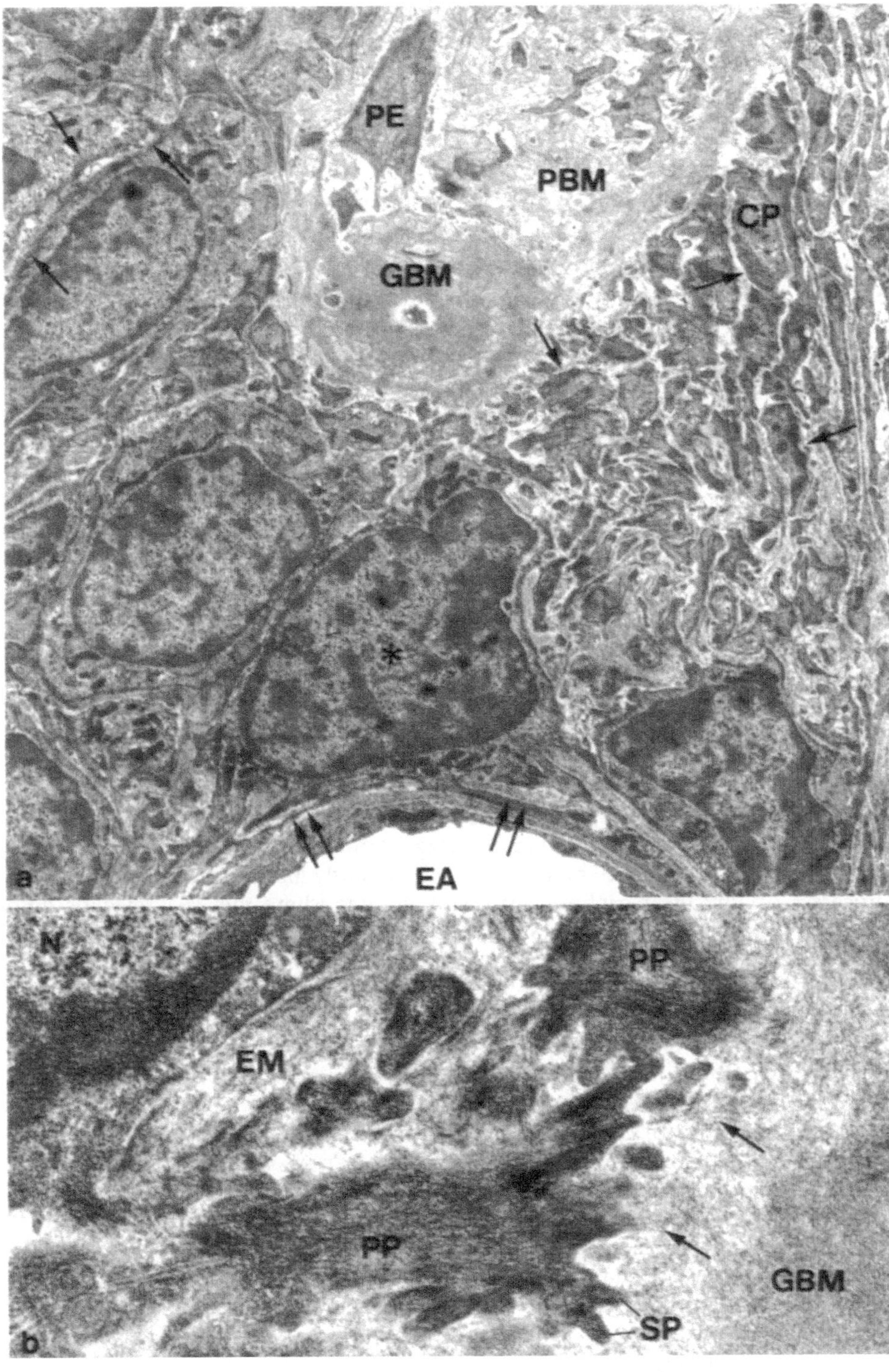
PE
PBM
CP
GBM
EA
a
N
PP
EM
PP
GBM
SP
b

connected to the EGM. Laterally, the EGM faces the renal interstitium. At this site, adrenergic nerves coming from the nerve plexus of the afferent arteriole pass by towards the efferent arteriole (Gorgas 1978, Luff 1992). Figures 27 to 30 show successive cross sections through the EGM.

3.3.1 EGM Cells

Cell Shape and Arrangement. The elongated EGM cells are highly branched cells consisting of a flattened, disc-shaped cell body, from which numerous cell processes arise (Spanidis et al. 1982). Two major categories of cell processes can be distinguished: large, thick "primary" processes emerging from the cell body and thin, comparably short "secondary" processes originating directly from the cell body and from the primary processes. The branching of primary processes into thin and short secondary processes is especially prominent at the periphery of the EGM, where these processes attach to the PBM (Figs. 28, 30b, 31).

The EGM cells are piled up in two to three layers parallel to the base of the macula densa (Fig. 11). In the vicinity of the arterioles, the layers adapt to the outer contours of these vessels and contain the EGM cells that partially surround the arterioles and become closely associated with smooth muscle cells (SMC) (Fig. 29). Discrimination between EGM cells and SMC may be difficult, but when tracing a cell in several parallel sections, clear-cut identification is possible (see also footnote on page 25).

The cell body and corresponding processes are always located within the same layer of the EGM (Fig. 26). The EGM cells send processes predominantly in one direction within an EGM layer. This was clearly demonstrated in a previous study in rabbit kidney by tracing individual EGM cells in serial sections and subsequent reconstruction (Spanidis et al. 1982). It was shown that the processes emerging from one side stayed together and projected in one direction. Thus, all these processes had close spatial relationships with each other, forming a "bundle" of processes.

Within the EGM field cell bodies are predominantly located in the center; the cell processes occupy the periphery. Especially at sites where the EGM extends onto Bowman's capsule, extensive regions of cell processes are found adhering to the PBM (Figs. 26, 29, 30).

Cytological Aspects. The EGM cells contain a large nucleus with deep invaginations of the nuclear membrane. Corresponding indentations of the cell membrane often protrude into these invaginations of the nuclear membrane, which results in a perplexing and extensive proximity of both membranes (Fig. 31). A Golgi field and a prominent centriole are located adjacent to the nucleus (Fig. 32). The cell body and the cytoplasm of the primary processes contain polysomes, lysosome-like inclusions, and mitochondria. Clathrin-coated micropinocytotic vesicles and few caveolae are present along the cell membrane of the cell body and the primary processes (Fig. 31).

In EGM cells the microfilament system is well developed and is best visualized in specimens treated with tannic acid. The microfilaments form a conspicuous layer directly beneath the cell membrane (Figs. 31, 32). As a whole, the microfilaments run longitudinally within the cell body and the primary processes and extend as prominent bundles into the secondary processes. Frequently, two secondary processes, projecting

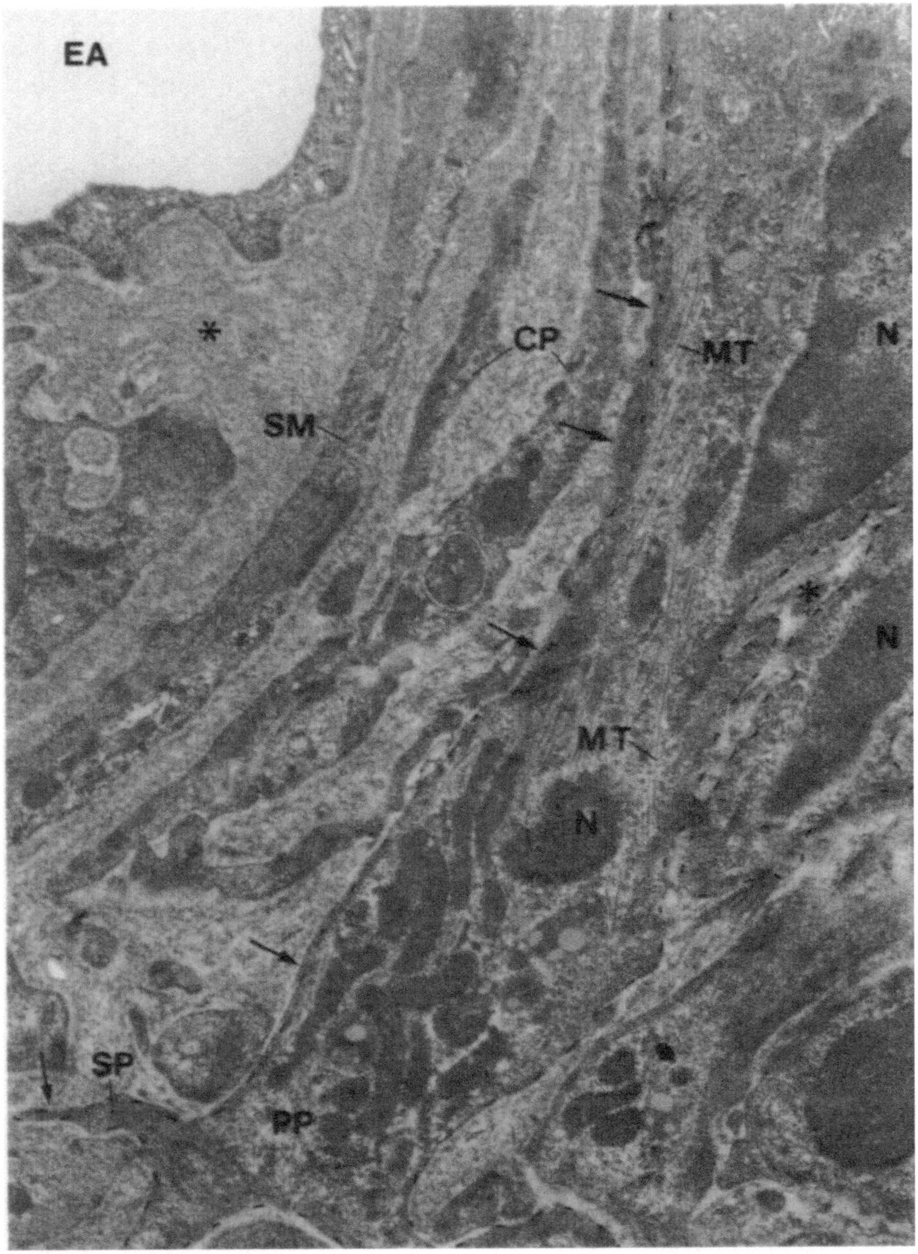

Fig. 31 Cross section through the vascular pole showing the substructure of EGM cells. Part of a cell body with its nucleus (*N*), as well as a primary process (*PP*) and secondary processes (*SP*) are seen. Microtubules (*MT*) extend from the cell body into the PP. Dense accumulations of microfilaments are located beneath the cell membrane and in SP (*arrows*). The plasma membrane projects into the cell body (*asterisk*). The *left upper corner* shows the EA with a continuous endothelium, under which is basement-membranelike material (*asterisk*), which is finally surrounded by a smooth muscle layer (*SM*). TEM, ×~22 800

Fig. 32. Cross section through the vascular pole, demonstrating the mechanical interconnection of EGM cells. Several EGM cells with microfilament assemblies adjacent to the cell membrane are seen. Individual microfilament bundles of one cell appear to have a microfilament bundle continuing in the same direction in the next cell (*thick arrows*). The interposed extracellular matrix has fine filaments (*thin arrows*). The matrix-mediated interconnections seem to be based on pairs of focal contacts formed by two neighboring cells. *C*, centriole; *CI*, cortical interstitium; *N*, nuclei of EGM cells. **TEM,** ×~9500

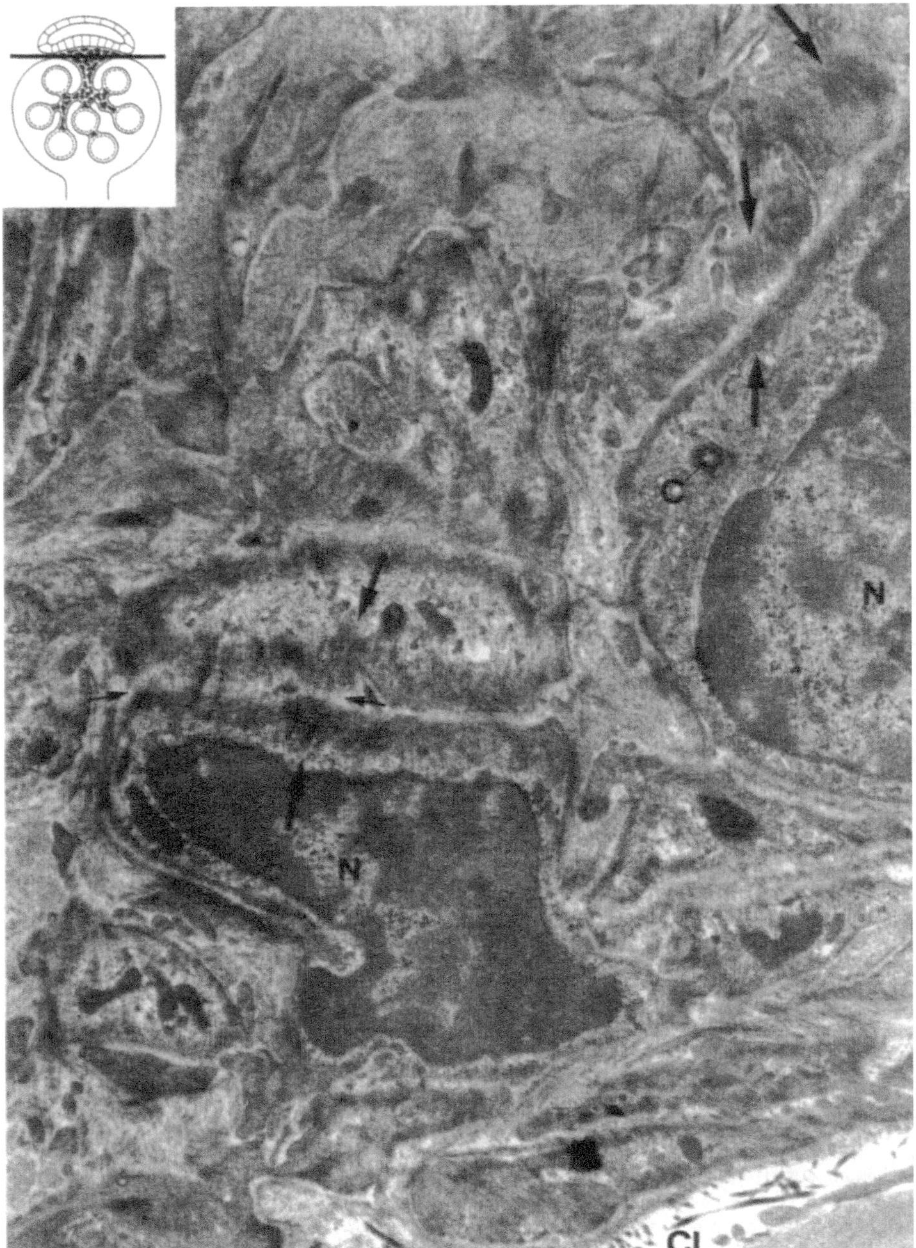
C
N
N
Cl

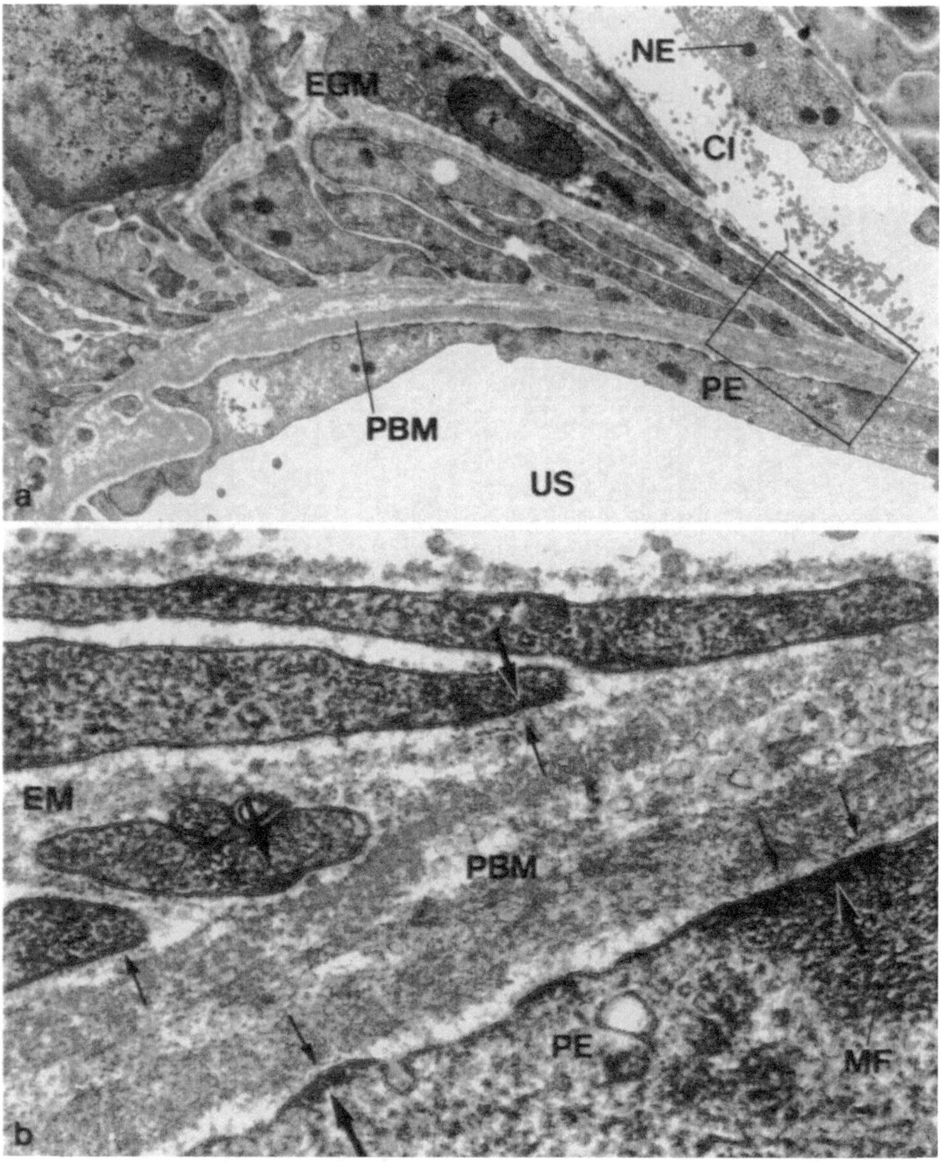

Fig. 33a,b. Interarteriolar section showing the attachment of the EGM to the PBM. **a** The peripheral part of the EGM is seen, which consists of cell processes extending towards and attaching to the PBM. **b** Higher magnification of area delineated in **a**. Cell-matrix connections are present near the end of the cell processes: the cell membrane is associated with electron-dense plaques on the cytoplasmic side (*large arrows*) and fine-filamentous material on the outside (*small arrows*). Similar cell-matrix contacts are also seen with parietal cells (*PE*). Note that the EGM matrix (*EM*) is continuous with the PBM. *CI*, cortical interstitium; *MF*, microfilament bundle; *NE*, nerve; TEM, **a** ×~12 800, **b** ×~81 000

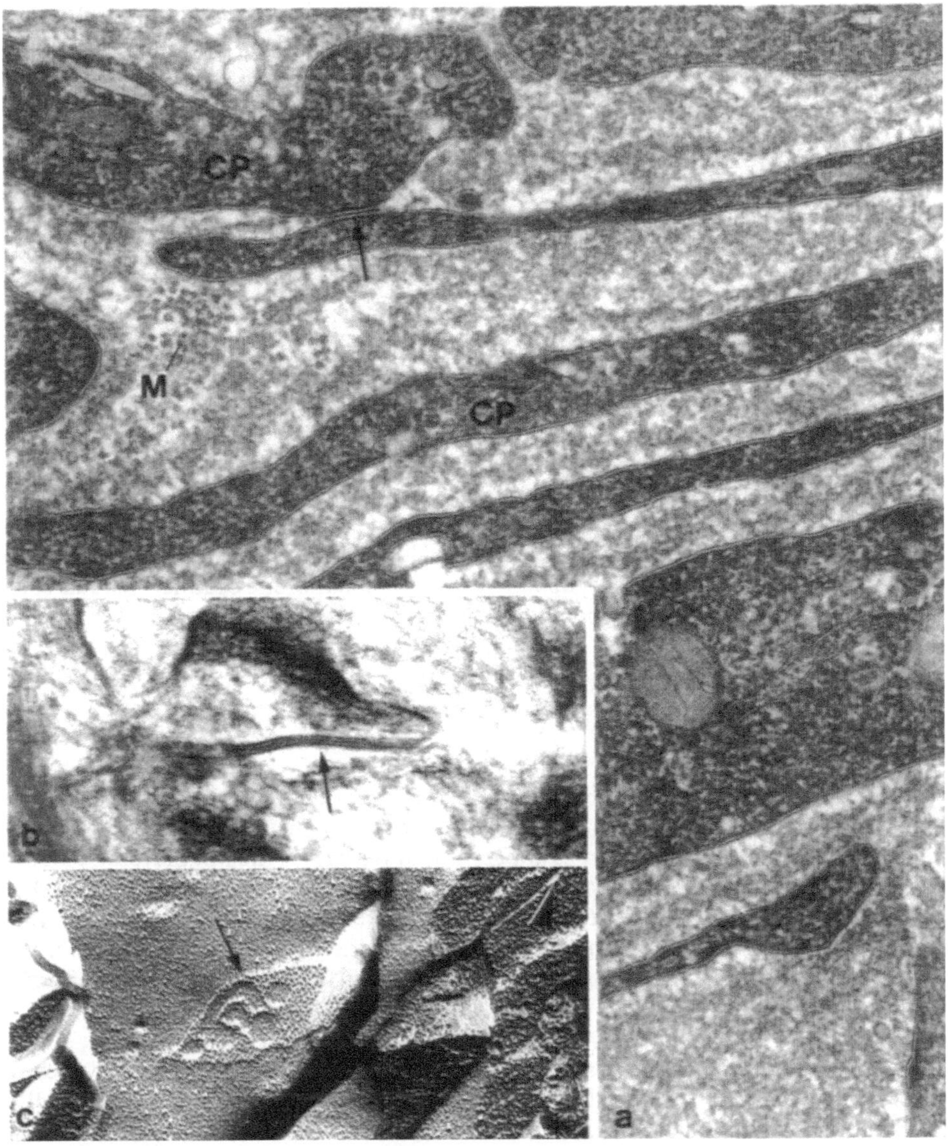

Fig. 34a–c. EGM cell-cell contacts and EGM matrix. **a** Several cell processes (*CP*) can be seen that are separated by matrix. The matrix is granular in appearance and varies in density. A lamina raralike layer is often seen between the cell membrane and more dense portions. In this figure microfibrils (*M*) are seen, but are not very common in EGM matrix; their tubular character is recognized in cross section. Cell processes may be interconnected by gap junctions (*arrows*), as seen in **a** and **b**, and in a freeze-fracture replica in **c**. TEM, **a** ×~71 400, **b** x~90 000, **c** 40 000; **b** and **c** from Taugner et al. (1978), with permission

from a primary process in opposite directions, form a pair that contains a continuous microfilament bundle extending from the tip of one process to the tip of the other. Microfilament bundles traversing the axis of a primary process in oblique directions are also frequently encountered. Attachment sites of microfilament bundles are seen along the cell membrane of the cell body, of primary processes, and also in the tips of the secondary processes (Figs. 30b, 33).

Longitudinally arranged microtubules extend from the cell body into the primary processes, but seem to be absent in the secondary processes (Fig. 31). Intermediate filaments are present throughout the EGM cells, as can be deduced from the immunostaining pattern for desmin and vimentin (Bachmann et al. 1983).

EGM cells are frequently involved in the formation of cell-cell contacts (Fig. 34). As can be seen best by freeze-fracture electron microscopy, the majority of the cell contacts are gap junctions (Pricam et al. 1974; Boll et al. 1975; Taugner et al. 1978). They are found in different locations: first, they interconnect neighboring EGM cells and, frequently, even processes of the same EGM cell; second, gap junctions connect EGM cells with mesangial cells, arteriolar smooth muscle cells, and granular cells. Thus, all smooth-muscle-like cells at the vascular pole are interconnected; gap junctions of EGM cells with macula densa cells do not exist. The extensive coupling of EGM cells by gap junctions was also visualized by immunohistochemical localization of connexin 43, a protein of the gap junctional transmembrane channel (Barajas et al. 1994). Desmosomes are occasionally encountered between the EGM cells.

3.3.2
EGM Matrix

The EGM cells are surrounded by extracellular matrix (EGM matrix) (Figs. 30, 33, 34). This matrix is not homogeneous, varying in electron density and texture. In many places, a loose, layered structure is apparent that resembles the multilayered parietal basement membrane rather than the intraglomerular mesangial matrix (Fig. 34). Unlike the mesangial matrix, tubular microfibrils (diameter: 12–15 nm) were only occasionally encountered. Immunocytochemical evidence has been provided for the presence of type IV collagen, fibronectin and laminin in the EGM matrix (Blanc-Brunat et al. 1989; unpublished results from our laboratory).

Cell-Matrix Relationships. The microfilaments of the EGM cells are frequently seen to terminate in dense plaques on the inner side of the plasma membrane (Figs. 30b, 33b). In obliquely sectioned cell membranes, bundles of microfilaments appear in close association with densely stained thin fibrils on the outside of the cell, which project in the same direction as the intracellular microfilaments (Fig. 32). This tandem arrangement of cytoplasmic microfilaments and extracellular fibrils is indicative of specialized contacts between cells and matrix (Burridge et al. 1988). Extensive formation of focal adhesions has been described in cultured mesangial cells (Woods and Couchman 1992; Petermann et al. 1993). Fibronectin, which is present in the EGM in a focal distribution, as shown by immunofluorescence (Blanc-Brunat et al. 1989), probably plays a role in these contacts. In favorable sections, microfilaments appeared to continue via extracellular fibrils with the microfilaments of a neighboring cell, suggesting a functional system interconnecting several cells. Thus, EGM cells and EGM matrix appear to act as a functional entity that allows for the mechanical interconnection of structures at the vascular pole.

3.3.3
EGM Relationships to Neighboring Structures

If we consider the EGM as the connecting medium between the structures of the vascular hilum, the anchorage of EGM cells to surrounding structures is of major interest. The EGM interconnects opposing parts of Bowman's capsule and the two arterioles (Figs. 26–29). The EGM processes affix themselves most frequently to the PBM. Since the EGM completely fills out the funnel-shaped space between the two arterioles, opposing parts of the funnel wall appear to be interconnected by the EGM cells (Fig. 26). In the most narrow parts of the funnel at the transition of the mesangium to the EGM, a single cell is sufficient to interconnect opposing parts of Bowman's capsule. As the funnel opens up, a chain of cells (interconnected among themselves by the matrix) is necessary to span opposite sides.

Relationships to Bowman's Capsule. The EGM is usually connected to Bowman's capsule by its numerous cell processes, but it can also be connected by the cell bodies themselves (Figs. 26, 31). The fixation of the cell processes on the PBM is generally arranged in a focal pattern. In grazing sections through Bowman's capsule it is evident that the fingerlike secondary processes are not strictly arranged in parallel to each other, but diverge in a roughly starlike pattern, making contact with the PBM (Fig. 30b). These contacts to the PBM are direct in the sense that the tips of the processes directly appose the outermost dense layer of the GBM or PBM without any other kind of extracellular matrix in between. In other places, cell bodies make elaborate contact with the basement membrane.

Abundant collagenous fibers and occasional fibrocytes are located adjacent to the EGM in the renal interstitium (Fig. 6). Along Bowman's capsule these collagenous fibers (identified by their thickness and periodic striation as type I and/or type III collagen) occasionally penetrate between the EGM and PBM. At this site, assemblages of extracellular granular vesicles are also found, sometimes forming extensive accumulations either between the EGM and the PBM, or between the outermost layers of the PBM (Figs. 5c, 6).

Relationships to the Arterioles. EGM cells also interconnect both arterioles. To understand the way this interconnection takes place, it is important to realize that the AA loses its smooth muscle layer earlier on the side towards the EGM axis than on the opposite side (Figs. 15a, 23, 35). The same is true at the EA (Figs. 21b, 36, 37). The EGM matrix merges with the endothelial basement membrane of these vessels (Fig. 35). EGM cell processes at this site are frequently arranged in parallel with the endothelium of the arterioles and are fixed to the basement membrane tangentially. Towards the periphery of the EGM, the EGM cells extend onto the smooth muscle layer of the arterioles (Fig. 36–38).

Relationships of the EGM and Glomerular Arterioles to the Macula Densa. The EGM and the juxtaglomerular portions of the glomerular arterioles have intimate

Fig. 35. Cross section through the vascular pole, showing the merging of the wall of the AA at the transition into the glomerulus. (Detail of Fig. 29). Along with the entry of the AA into the glomerulus, the smooth muscle cells (*SM*) towards the center of the EGM disappear and are replaced by EGM cells. The basement membrane of the AA (marked by a *broken line*) deviates from the endothelium and merges with the EGM matrix (*large arrows*). EGM cell processes (*CP*) running in various directions contain microfilament bundles. The beginning of the endothelial fenestration is indicated (*small arrows*). TEM, ×~13 700

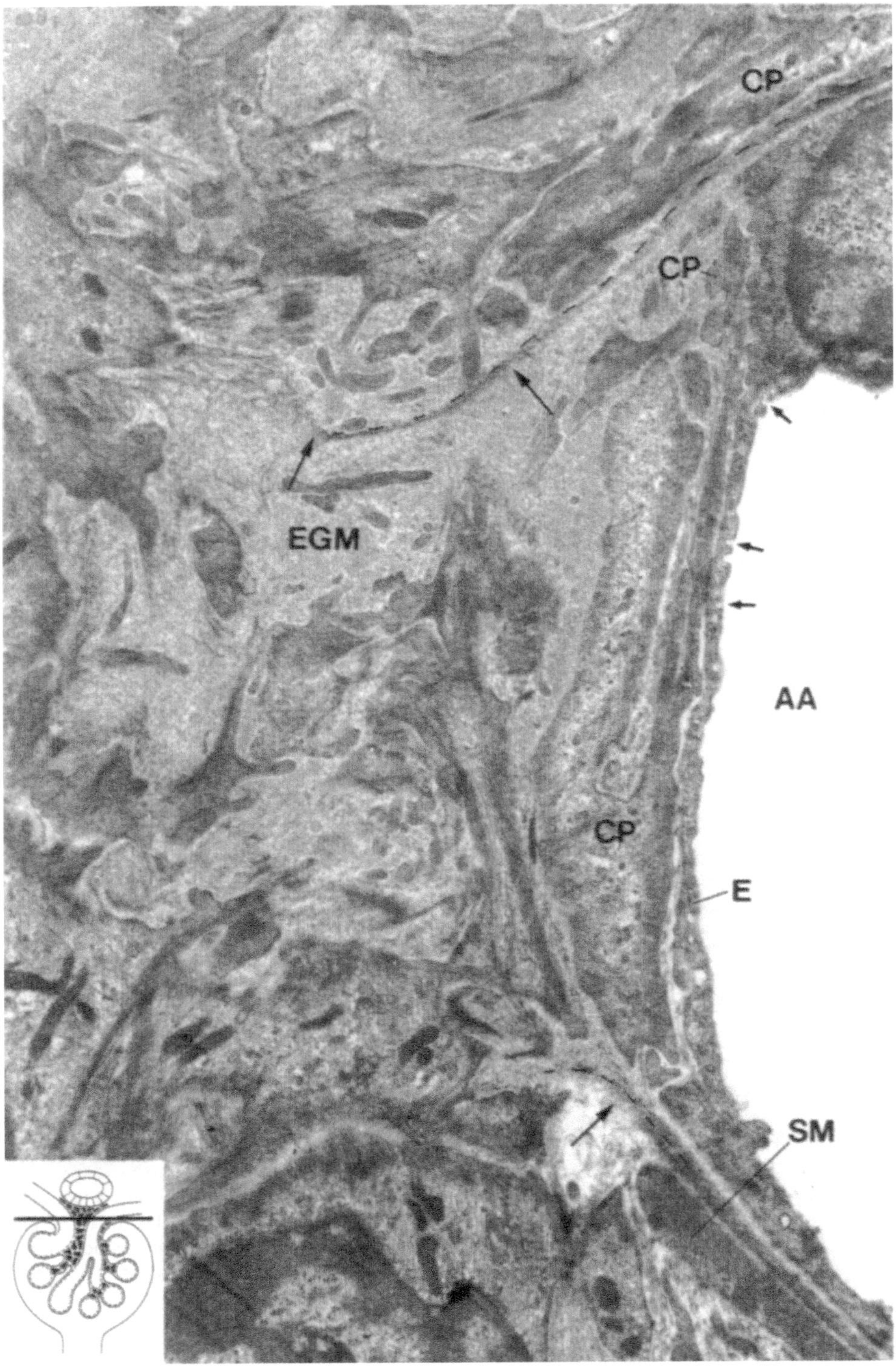
CP
CP
EGM
AA
CP
E
SM

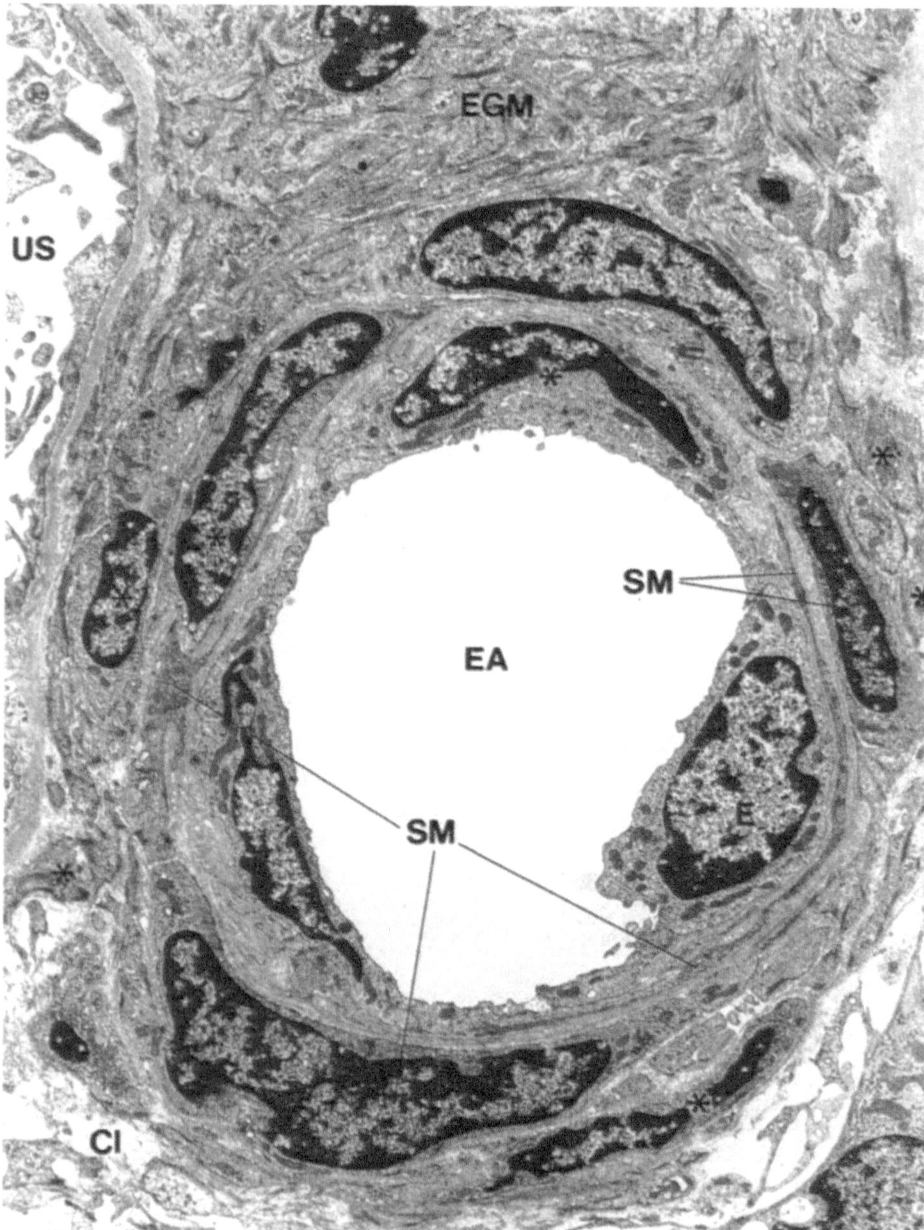

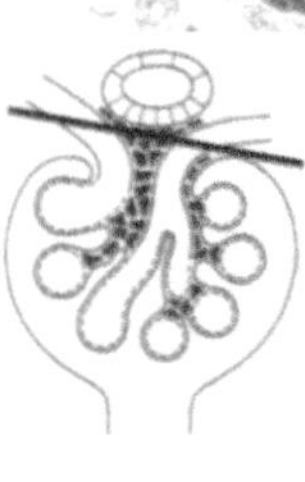

Fig. 36. Cross section through the vascular pole, showing the association of the EA with the EGM. Towards the EA the EGM cells adapt to the outer arteriolar surface. The EA is surrounded by two layers: an incomplete inner smooth muscle layer (*SM*), followed by an outer layer of EGM cells. *US*, urinary space; *CI*, cortical interstitium. TEM, ×~6500

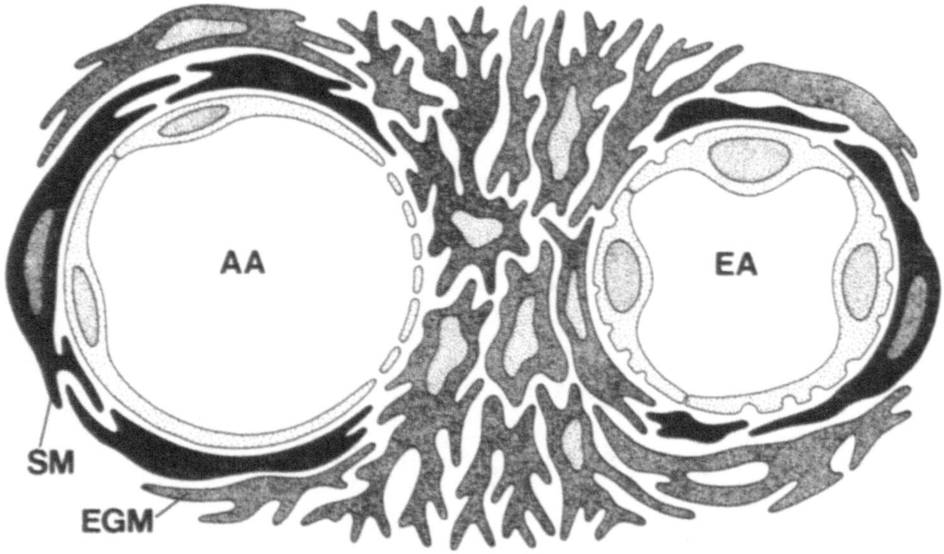

Fig. 37. Schematic cross section through the vascular pole just above Bowman's capsule. Afferent and efferent arterioles (*AA*, *EA*) are cut traversely. Extraglomerular mesangial cells (*EGM*) are shown in moderate gray, smooth muscle cells in dark gray, and endothelial cells in light gray. Note differences in the wall of AA and EA: The AA already displays endothelial fenestration on the side facing the EGM. Conversely, the EA has a continuous endothelium with many cell bodies. In both AA and EA the smooth muscle layer (*SM*) is not complete: towards the center of the EGM the SM cells are replaced by EGM cells

relationships to the macula densa (MD). These relationships are established early during glomerular development; in the S-shaped stage of the glomerular anlage, the prospective MD cells already abut to the prospective mesangial/extraglomerular cells (Saxén 1987; Fischer et al. 1995). In the adult the MD comprises some 20–30 specialized cells within the end portion of the cortical thick ascending limb. A short distance after the MD (after a postmacular segment) the TAL transforms into the distal convoluted tubule (Kriz and Kaissling 1992).

The basal surface of the MD attaches to the EGM; this contact is obligatory and is the only consistent relationship the MD has with the vascular pole structures. In addition, the MD is in contact with juxtaglomerular portions of the afferent and/or efferent arterioles. However, these relationships are variable and may be totally absent (Christensen and Bohle 1978). As seen by electron microscopy, EGM cells, often represented by only a thin cell layer, penetrate between the base of the MD and the muscle layer of glomerular arterioles in most places (Barajas 1970) (Fig. 13). Thus, at those sites no direct contact is established but, judged simply on the basis of light microscopy, direct contact would have been expected. Moreover, discrepancies in the literature concerning the topographical relationships of the MD with the AA and EA have emerged because the MD frequently has not been clearly distinguished from adjacent portions of the distal tubule (upstream and/or downstream). The situation is complicated by the fact that cells adjacent to the MD (perimacular cells), as well as individual cells scattered in the TAL epithelium distant from the MD, share structural

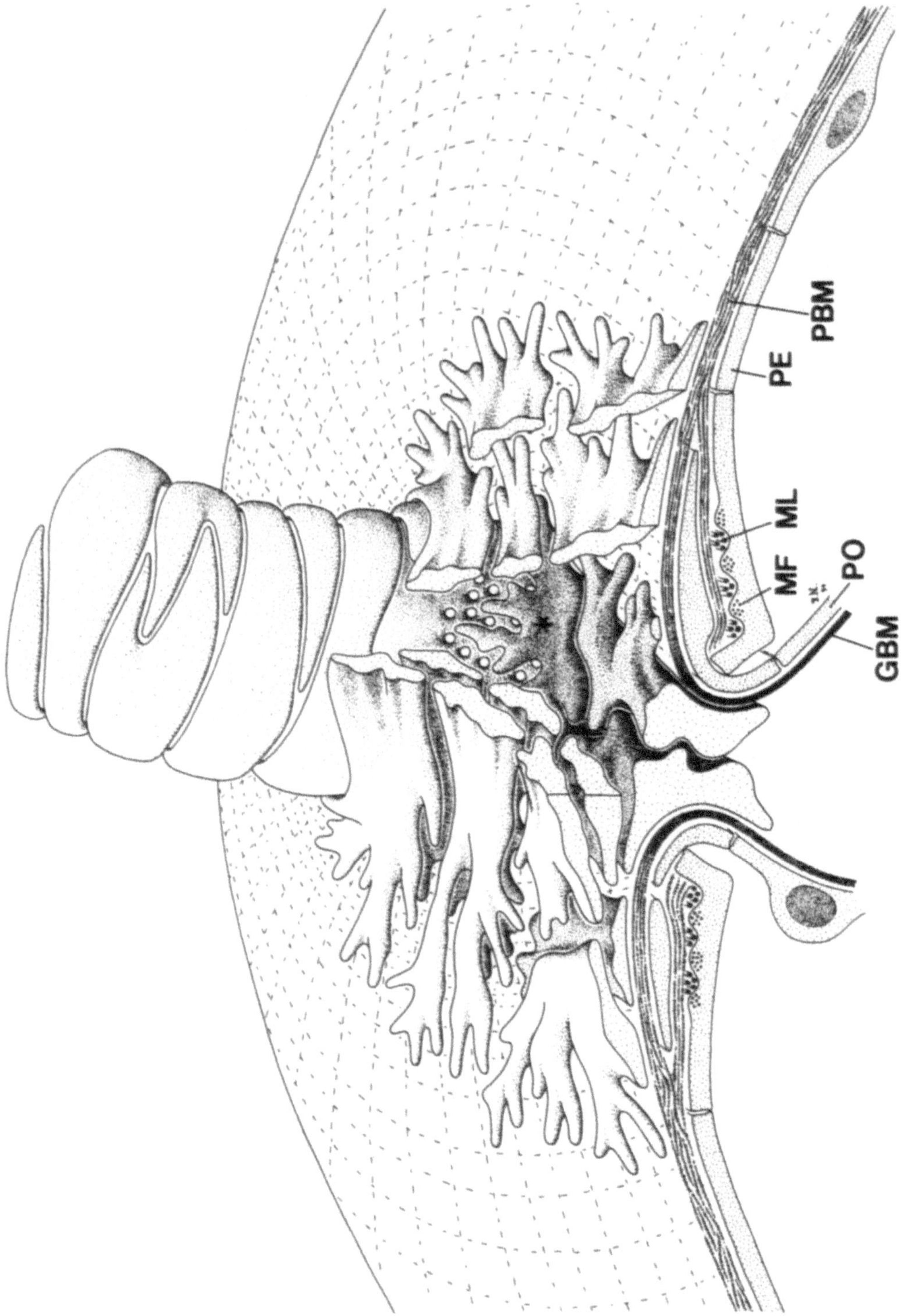
PBM
PE
ML
MF
PO
GBM

and immunocytochemical features with the MD cells proper [lack of Tamm-Horsfall glycoprotein and selective expression of neuronal nitric oxide synthase (Briggs and Schnermann 1987; Bachmann et al. 1995)]. These cells may directly (i.e., only separated by extracellular matrices) contact glomerular arterioles, the cortical interstitium or other nephron segments. Direct contact between granular cells of the AA and the MD or perimacular cells is only occasionally encountered (Gorgas 1978; Spanidis et al. 1982). However, there are additional contacts distant from the vascular pole between the EA and the TAL upstream to the MD (Barajas and Latta 1963; Faarup 1965). The functional relevance of these relationships is fully unknown and, unlike the MD, no special elaboration of the interface between these structures has been established. In conclusion, the only consistent relationships of the basal surface of the MD are established with the EGM.

The contact zone between the MD and the EGM exhibits several features that distinguish this region from the rest of the TAL (Fig. 26). The extracellular matrix layer beneath the MD cells may be smooth, showing up as the usual basement membrane, or it may be irregular, showing deep indentations towards the EGM (Bonsib 1986). The matrix layer of the MD abuts the EGM matrix. MD cell-to-matrix attachment at this site is suggested to be mediated by $\alpha v\beta 6$ integrin, a fibronectin-binding heterodimer (Busk et al. 1992). In kidney tissue, $\beta 6$ (found only in epithelial tissues) is specifically expressed in MD cells (Breuss et al. 1993). It is concluded that $\alpha v\beta 6$ is involved in the adhesion of MD cells to fibronectin of the EGM matrix (Blanc-Brunat et al. 1989).

3.3.4 Glomerular Stalk

The term "glomerular stalk" is widely used in the sense that the glomerular tuft has a stem piece, by which the tuft is anchored to the glomerular vascular pole; this stem piece is thought to contain the terminal and the beginning portions of both glomerular arterioles. Actually, a stalk in this definition is extremely short.

The two glomerular arterioles approach the vascular pole from opposite directions (actually the EA withdraws from the pole); thus, they approach the glomerular globe in a tangential way. They come nearest to each other at the level of the EGM. Here they make a sharp bend and enter the tuft in a more radial direction. However, after this

Fig. 38. Three-dimensional scheme demonstrating the relationships of structures at the vascular pole. One-half of the pole containing the afferent arterioles (*AA*) is shown. The reflection of Bowman's capsule with the transition from parietal epithelium (*PE*) to visceral (podocyte) epithelium (*PO*) and the transition from the parietal basement membrane (*PBM*) to the GBM border the opening of Bowman's capsule. On the *left side*, the branched EGM cells are seen; on the *right side* layers of EGM cells are partly removed to show deeper structures. At the AA, the deepest EGM cell (*asterisk*) attaches by its processes to the fenestrated endothelium. The smooth muscle cells form rings around the AA; EGM cells and SM cells may overlap. EGM cells of deep and superficial layers attach to the PBM by their processes. The interarteriolar section plane shows several details of the wall structure of Bowman's capsule: Circular ridges of parietal cells (*PE*) contain intracellular microfilament bundles (*MF*). In the furrows between the ridges extracellular "microligaments" (*ML*) run in parallel. Note that the first parietal cells or winglike processes are located between the dense layers within the PBM. *PO*, outermost podocyte

bend only the EA has a trunk piece (the intraglomerular segment of the EA), whereas the AA immediately divides into its first branches; thus, an afferent trunk segment running in a radial direction is generally lacking.

Within the EGM both arterioles lie near each other. A cross section through the vascular pole may cut both arterioles traversely, giving them round profiles, surrounded by extraglomerular mesangial cells (towards the center) and smooth muscle cells (towards the periphery). This situation is shown in the schematic drawings in Fig. 37. Slightly deeper and thus still at the beginning of the stalk, the AA starts to branch. Immediately beneath the level of the glomerular entrance, the trunk segment of the EA is lying opposite the branching portion of the AA. Therefore, the glomerular stalk does not really contain trunk segments of both arterioles, but contains the branching point of the AA and the trunk segment of the EA, enclosed together by the glomerular basement membrane. This situation is shown in Fig. 19. Therefore, a true stalk containing the trunks of both arterioles encompassed together by the GBM is never really developed. The tuft has no clearly discernible stalk, but is anchored very intimately to the EGM. The branching portion of the AA has a certain thickness of 6–7 μm; this distance may actually be taken as the length of the stalk.

4 Discussion

The discussion will address four major points. First, which structures and mechanisms are responsible for stabilization of the vascular pole against the expanding forces of hydrostatic pressure gradients. Second, what role does the arrangement of structures play at the vascular pole for regulating glomerular blood flow and filtration. Third, what are the relationships between the vascular pole structures and the macula densa (MD). Fourth, does the morphology favor the idea of considerable fluid leakage through the glomerular stalk into the cortical interstitium.

4.1 Stabilization of the Vascular Pole

4.1.1 Challenges/Pressure Gradients

A discussion of the stabilizing mechanisms has to begin with an analysis of the mechanical challenges to which the vascular pole is exposed. The hydrostatic pressures in the afferent arteriole (AA) and efferent arteriole (EA) are high, ranging in the order of 70–50 mm Hg, and may rise to much higher values in hypertension. These pressures create pressure gradients across several interfaces.

Like elsewhere in the body, pressure gradients occur across the arteriolar walls. However, since the complete muscle layer that provides circular support for the AA and EA gradually disappears (or is gradually reestablished), structures other than simply the arteriolar wall come into play, counteracting transmural pressure gradients. Since the AA not only loses its smooth muscle layer but also its basement membrane (first towards the extraglomerular mesangium, EGM) concurrent with the appearance of endothelial fenestrations, the pressure barrier as a whole disappears; a correspon-ding situation is seen at the EA, where a pressure barrier is gradually developed. Consequently, we have to assume that the high arteriolar pressure spreads to some extent into the EGM and the mesangium of the glomerular stalk (comparable to the situation in glomerular capillaries where the capillary pressure is carried to the mesangial space proper (Kriz et al. 1990a,b).

Figure 39 schematically summarizes pressure gradients at the vascular pole. Arrow A in Fig. 39 depicts the pressure gradient across the wall of the glomerular stalk towards Bowman's space. For the time being, we can assume that there are no remarkable hydrostatic pressure differences among the compartments inside the stalk, i.e., between the afferent lumen, the mesangium and the efferent lumen (a detailed analysis

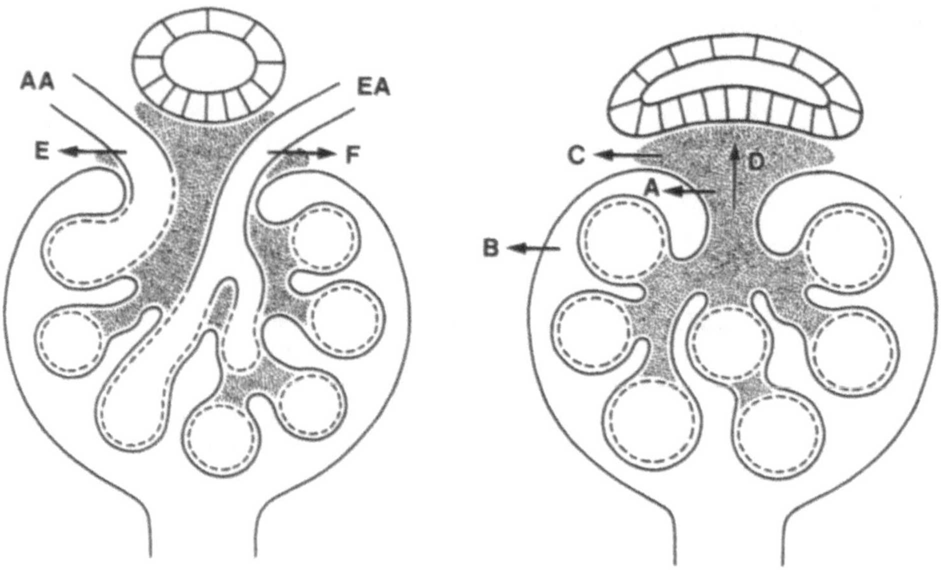

Fig. 39. Schematics of trans- and interarteriolar sections, emphasizing the vascular pole to demonstrate hydrostatic pressure gradients between the various compartments within the vascular pole. Gradients between capillary lumen and urinary space, capillary lumen and mesangium, and between mesangium and urinary space are not included. See text for further explanation

is given below). Arrow B depicts the pressure gradient across Bowman's capsule between Bowman's space and the cortical interstitium. Arrows C and D depict the pressure gradient between the interior of the EGM towards the cortical interstitium and the base of the macula densa, respectively. The pressure differences across the walls of both arterioles (arrows E and F) are also shown.

In addition, the pressure gradients at the vascular pole may act in a conjoint way. Together with the entity of the glomerular capillaries, the glomerular arterioles form a hairpin loop. The AA enters the tuft, falling apart into the capillary network, out of which the EA develops and leaves the glomerulus in the vicinity of the AA. Imagine a simple garden hose that makes a bend. Considerable counterforces are necessary to maintain the bend when the hose is pressurized. If those forces are lacking, the hose will simply straighten out. This effect derives from the fact that in a loop the forces acting on the convex wall part of a loop are always greater than those acting on the concave side.

As can be seen in Fig. 39, the pressure barrier for the high "interstitial" pressure in the glomerular stalk (as in the intraglomerular mesangium) is the GBM; no other structure exists that could effectively serve as a pressure barrier. The GBM then reflects to pass over into the PBM, thereby establishing the opening of the glomerular entrance. Thus, the border of this reflection is the outermost structure exposed to the pressure gradients arising from the high hydrostatic pressures inside the vascular pole. Towards the MD as well as towards the cortical interstitium, the layered arrangement of the EGM itself appears to represent a staggered pressure barrier, which apparently allows a gradual decrease in these pressures.

In addition, there is a pressure gradient - even if a much smaller one - across Bowman's capsule, leading to expansile forces and protecting the urinary space from collapse. These expansile forces give rise to tension lines within the wall of Bowman's capsule, running virtually in all directions (meridional, equatorial and any other direction). Together with the PBM, the contractile apparatus of parietal epithelial cells is believed to develop counterforces against distension of Bowman's capsule (Lebrun et al. 1992; Mbassa et al. 1988). In the present context, the *meridional counterforces* are of interest since they might contribute to dilation of the opening in Bowman's capsule.

4.1.2
Counterforces

The distending forces acting on the vascular pole must be balanced by counterforces, which may be subdivided into: (a) circular forces developed by the walls of both arterioles and by the wall of Bowman's capsule encircling the opening; (b) interconnecting forces developed by the EGM between both arterioles and between opposite portions of the wall of the opening; (c) occluding forces developed by the EGM between the stalk on one side and the MD and cortical interstitium on the other.

Circular Forces. Circular forces are developed by the walls of both glomerular arterioles, as well as by Bowman's capsule around the entire vascular pole. The AA, when approaching the glomerular entrance, is a proper arteriole with a continuous endothelium, a basement membrane and a complete layer of smooth muscle cells. At sites where smooth muscle cells are replaced by granular cells near the vascular pole, granular cells complete the ring of contractile cells surrounding the endothelial tube. In vivo observation has shown that this most distal part of the AA is still capable of constriction (Steinhausen et al. 1989, 1990; Takenaka et al. 1994; Sanchez-Ferrer et al. 1989; Moore and Casellas 1990; Kimura et al. 1990; Edwards 1983; Wilson 1986). Thus, it is evident that the AA is able to develop circular forces up to its entry into the glomerulus. The same holds true for the EA as soon as it is established as a proper arteriole; after leaving the glomerulus, a complete layer of contractile cells (EGM cells and smooth muscle cells) surrounds the vessel (Fretschner et al. 1990).

Circular forces are also developed around the vascular pole. The skeletal elements encircling the opening of Bowman's capsule are the GBM and the beginning portions of the PBM. The short glomerular stalk is encompassed by the GBM. Outwards the GBM funnel opens up; finally, the GBM reflects and transforms into the PBM. Along this reflection the podocytes pass over into parietal cells.

The glomerular stalk is covered by podocytes, although they attach to it by coarse processes or by cell bodies; an interdigitating pattern of foot processes is poorly developed at this site. The basal cytoplasm of these podocytes generally contains prominent microfilament bundles arranged in a circular manner around the glomerular stalk. Several cells are necessary to establish a complete ring around the stalk. Since force-transducing structures between podocytes are not evident, the mechanical continuity is apparently provided by the GBM. The mechanical fixation of podocyte foot processes to the GBM is mediated by $\alpha 3\beta 1$ integrins (Cybulsky et al. 1992). This is probably the case for the podocytes attached to the stalk as well. Thus, the GBM together with the podocytes should be able to develop wall tension, i.e., centripetally directed circular forces counteracting the expanding forces originating from pressure differences across this wall.

The mechanical situation changes when the GBM funnel opens up. The angle between the wall and plane of the circular structures, which is roughly 90° at the stalk, grows larger. Consequently, the situation becomes more and more unfavorable for structures to develop effective circular forces. Nevertheless, there are prominent circular structures that are obviously adapted to the specific mechanical situation (Mbassa et al. 1988). Microfilament bundles in the basal ridges of the first row of parietal cells run in a circular direction around the pole. They are separated from each other by furrows indenting into the cell, which are filled with circularly running extracellular fibers, called microligaments. Thus, a circular skeletal element (the microligaments) is combined with a circular contractile element, establishing a unit that appears to be able to develop circular forces.

Interconnecting Forces. Interconnecting forces may be developed by structures that bridge opposing parts of Bowman's capsule in addition to fixing the two arterioles to each other. Deep in the funnel, i.e., in the glomerular stalk, a single mesangial cell may interconnect the opposite sites. As the funnel opens up, a more elaborate system of interconnecting cells (EGM cells) is necessary to bridge the gap. Since there are no force-transducing intercellular junctions between EGM cells (desmosomes have only been found occasionally), the extracellular matrix should conjoin the cells to a block to which the walls of the funnel adhere. A trajectorial arrangement of microfilament bundles – presumably indicating tension lines – is regularly seen to jump from cell to cell: Microfilament bundles terminate at the plasma membrane of one cell and another bundle continues in the same direction in the next cell. The insertion of the microfilaments in the plasma membrane is reminiscent of focal adhesions (Burridge et al. 1988), where intracellular actin filaments are fixed by membrane-spanning integrins to the extracellular matrix. Thus, the connecting medium between the EGM cells is the matrix allowing force transduction from one cell to the next. Finally, this trajectorial system attaches to the wall of the opening by adhesion of the EGM cells to the basement membrane.

The connection of EGM cells to the arterioles is somewhat different. EGM cells, when approaching the AA or EA, intermingle with smooth muscle cells of the arteriolar walls, thereby encircling the vessel, at least partially, and adhering tangentially to its basement membrane.

Since most of the trajectorial microfilament systems interconnect opposing sites of the opening, they run in planes perpendicularly to the vertical axis of the funnel. Within a given plane – which may be imagined as a disc of one cell layer – the bundles extend in all directions. This arrangement may function as a spiderlike clamp, which can develop centripetally directed forces, preventing expansion of the glomerular opening.

Occluding Forces Counteracting the Pressure Gradient Across the EGM. Assuming that the interstitial pressure in the mesangium of the glomerular stalk is high, a considerable pressure drop is present across the EGM towards the base of the MD and the cortical interstitium. Thus, the EGM as a whole acts as a pressure barrier. This barrier consists of several layers of cells and matrix. This arrangement appears to be consistent with the view that the pressure drop across the EGM is gradual, decreasing stepwise from layer to layer.

An important aspect is that EGM cells are exposed to mechanical stress in two different directions. Interconnecting the borders of the glomerular opening, EGM cells are exposed to tension in a longitudinal (meridional) direction. The pressure gradient towards the MD tends to bend EGM cells towards this epithelium. Theoretically, the vectorial addition of both forces leads to curved tension lines. EGM cell layers clearly form arches (as seen in the interarteriolar section plane). Whether this is due to the dual strain to which the cells are exposed is unknown.

The site of interference of the various forces discussed in this chapter is the wall of the opening in Bowman's capsule. This wall shows several structural peculiarities, which may be interpreted to reflect this particular mechanical challenge. The differently shaped processes of parietal cells at this site may be seen as a result of a specific mechanical strain. Another aspect is the membrane-bound granular vesicles consistently found between the layers of the PBM around the opening. Vesicles at this site have previously been described (Johnston et al. 1973). In an identical location, processes of parietal cells as well as large accumulations of cell debris are seen. Therefore, we believe that the vesicles result from the disintegration of those processes (or of entire parietal cells). Vesicles of this type are generally considered to reflect cell destruction at sites not accessible to phagocytosis by macrophages or other mechanisms meant to clean up cell debris (Waugh et al. 1967; Johnston et al. 1973; Burkholder et al. 1973; Olsen et al. 1974). Therefore, the vesicles appear to indicate that there is considerable cell destruction and renewal (turnover) among parietal cells at the vascular pole. We may speculate that the different mechanical challenges at this site of interference of several expansile forces and corresponding counterforces may underlie these phenomena.

4.2 Regulation of Glomerular Blood Flow and Filtration

Our discussion of the hemodynamic relevance of the structures at the glomerular entrance will mainly address two questions. First, what are the hemodynamic consequences that arise from the narrow association of the AA and EA in the glomerular stalk and, second, what is the role of the outflow segment of the EA (EA_{OS}) in the regulation of glomerular filtration.

4.2.1 The Assemblage of AA and EA

The narrow juxtaposition of AA and EA at the glomerular entrance represents a very particular arrangement of unknown relevance. In Figs. 40 and 41, the essential features of this arrangement are summarized. We will first discuss this issue in the context of the entire glomerular tuft (Fig. 39).

The terminal portion of the AA, its first branches, the glomerular capillaries, the intraglomerular portion of the EA (EA_{OSi}) and the mesangium are contained in a common compartment that is enclosed by the GBM. The GBM represents the pressure

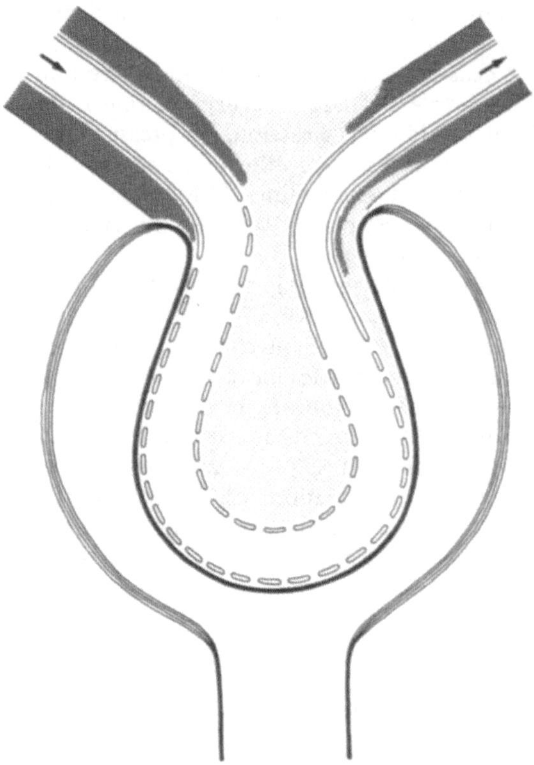

Fig. 40. A simplified diagram of the glomerulus. The network of glomerular capillaries is reduced to a single capillary loop. At the vascular pole, the GBM (black line) is reflected into the multilayered parietal basement membrane. The epithelial layers are not shown. Smooth muscle tissue is shown in dark gray, mesangium (*M*) and the EGM in light gray. The AA loses its smooth muscle layer at the entrance of the glomerular tuft and abruptly splits into porous glomerular capillaries. The endothelium peripherally faces the GBM and centrally the mesangium. The EA is established inside the glomerulus and passes through the mesangium (it is fully surrounded by a mesangial layer!) and the EGM. Along this course, smooth muscle cells gradually replace mesangial and EGM cells. Since there is no distinct hydraulic barrier between glomerular capillaries and the mesangium, the hydraulic pressure in the mesangium is probably in the same range as the capillary pressure. The EGM is expected to mediate the pressure drop towards the cortical interstitium. The functional implications of this arrangement are discussed in the text

barrier of the glomerular tuft separating a high-pressure compartment from the low-pressure compartment of Bowman's space (Drumond and Deen 1994). Within the high-pressure compartment the hydraulic pressures in the various subcompartments – afferent arteriole (P_{AA}), capillaries (P_{GC}), efferent arteriole (P_{EA}), mesangium (P_M) – are probably very similar, as large pressure differences are unlikely. The high pressure in any of them derives from the hydraulic pressure in the AA. As is generally agreed, the axial hydraulic pressure drop along the glomerular capillaries is small, amounting to only a few mmHg (Allison et al. 1974). Since there is no hydraulic pressure barrier between glomerular capillaries and the mesangium (the AA when approaching the glomerular entrance has an endothelium with large, open pores and loses its basement membrane), the hydraulic pressure in the mesangium, including the mesangium in

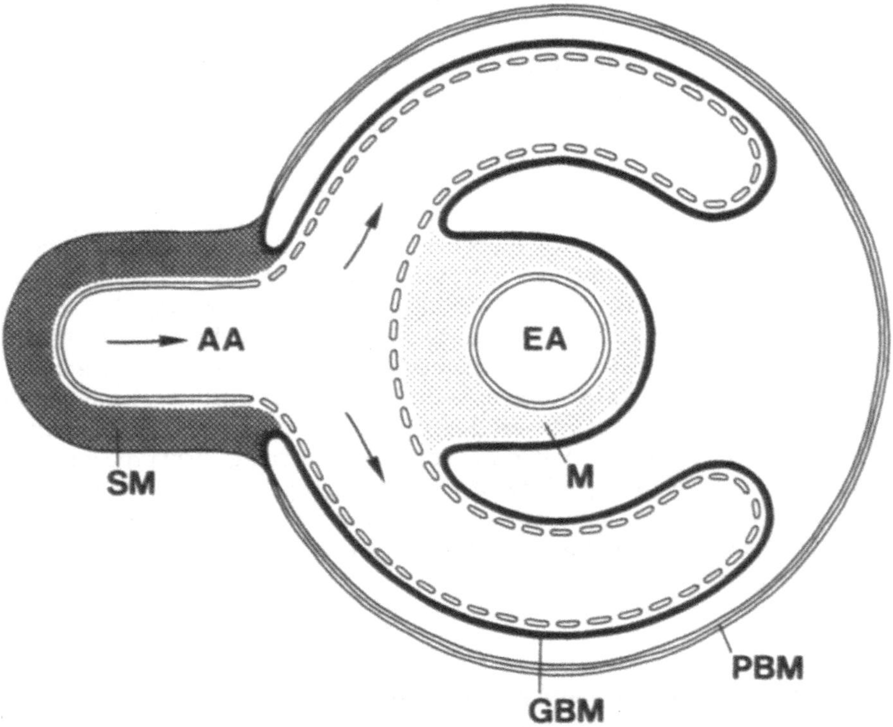

Fig. 41. Schematic of a cross section through the vascular pole, showing the spatial relationships of the AA and EA within the glomerular stalk (corresponding to the situation in Fig. 19). Immediately after its entry into the glomerulus, the AA splits into wide capillary branches with open endothelial pores. The branching point of the AA has a narrow spatial association with the outflow segment of the EA. The outflow segment is enclosed together with the AA in a common compartment bordered by the GBM. The EA is completely surrounded by a layer of mesangial tissue (*M*) and is separated from the AA only by this layer; there is no basement membrane between AA and EA. *SM*, smooth muscle layer

the glomerular stalk, is probably more or less identical to the capillary pressure (Lemley et al. 1992; Kriz et al. 1990a,b).

The situation in which the two arterioles would run in parallel over a long distance within the stalk does not develop. Since the AA already branches at the entrance level, the EA_{OSi} is generally most narrowly associated with the branching point of the AA. At this site the EA_{OSi} is fully surrounded by a mesangial layer, which directly faces the fenestrated endothelium of the branching point of the AA. The AA, its first branches, and the EA_{OSi} with its circular mesangial layer are encompassed together by the GBM, thus enclosed in a common compartment, which represents the glomerular stalk.

The pressure inside this compartment (P_M) may readily be expected to be fairly identical to P_{AA}. Because the EA at this site is only made up by an endothelial tube, the width of this vessel will largely depend on the balance between intraluminal (P_{EA}) and the surrounding mesangial hydraulic pressure (P_M). Since P_M – as just discussed – is probably identical to P_{AA}, the resistance of the EA at this site will critically depend on the ratio between P_{AA} and P_{EA}.

The EA_{OSi} changes in structure when approaching the vascular pole. First, the fenestrated endothelium is replaced by a continuous type and, second, the EA_{OSi} becomes fully incorporated into the mesangium. This leads to the situation shown in a schematic cross section through the glomerular stalk in Fig. 40.

At the beginning of the EA_{OSi} (because of the minor axial pressure drop along glomerular capillaries) the intraluminal pressure should not be much lower than P_{AA}. However, the pressure drop inside the EA_{OSi} from the site of confluence up to the exit from the glomerulus is probably considerable. In contrast, a pressure drop in the surrounding mesangium alongside the EA_{OSi} is improbable. Consequently, in vessel portions approaching the glomerular exit, the intravascular hydraulic pressure should become gradually lower than the surrounding pressure in the mesangial space. The narrow lumen of the EA_{OSi} just before leaving the glomerular stalk (as is usually seen) might be the result of a balance between comparably high pressure in the mesangium and the intravascular pressure, together with the inherent structural stability of the vessel wall. The embedment of this vessel segment into a mesangial layer represents a specific structural support with unknown relevance. This arrangement probably limits narrowing as well as widening of the efferent lumen, suggesting that the vessel can neither be fully closed nor extremely expanded. From this point of view the EA_{OSi} represents some sort of a fixed resistance which, within narrow limits, may passively adapt to the AA pressure.

4.2.2
The Role of the EA_{OS} in the Regulation of Glomerular Filtration

It is well established that the EA is a major site of regulation of glomerular filtration (Navar et al. 1996; Davis 1991; Ren et al. 1996; Endlich et al. 1996; Häberle 1988). Changes in efferent resistance have a decisive impact on filtration pressure and thus on the effectiveness of filtration. An increase in efferent resistance will result in an increase of filtration pressure followed – in a situation of filtration disequilibrium – by an increase of filtration fraction. In contrast, a decrease in efferent resistance will lead to a decrease in filtration fraction. It is generally assumed that the changes in efferent resistance are mediated by vasoactive substances acting on the EA outside the glomerulus (Navar et al. 1996).

The current understanding of glomerular blood flow and filtration regulation is based on the tuning of the interplay between afferent and efferent resistances, together with the regulation of the ultrafiltration coefficient, K_f. The way K_f is regulated is unknown. It is puzzling that this model of glomerular blood flow and filtration regulation does not include any signal coming from the glomerulus itself. Blood flow through the glomerulus and filtration is generally described as being exclusively regulated on the basis of signals whose regulatory loops are established outside the glomerulus. This is quite unsatisfactory for such a fundamental system. Therefore, it appears reasonable to ask whether the structural organization would offer possibilities to measure relevant parameters of glomerular blood flow and filtration dynamics inside the glomerulus. There are two likely parameters: pressure and flow.

Pressure. Blood pressure is generally believed to be sensed as tension of the vessel wall generated by the pressure difference across the wall. As a result of widening of the vessel, smooth muscle cells are stretched, followed by an increase in tension that is directly translated into an increase in contractile tone. This behavior of vascular smooth muscle cells is generally known as "myogenic response" (Harder et al. 1987; Aukland 1989). In the kidney, only the preglomerular vessels respond in this way; there is no evidence that efferent arterioles possess myogenic responsiveness to changes in hydrostatic pressure (Edwards 1983; Hayashi et al. 1989; Steinhausen et al. 1989; Fleming et al. 1987; Loutzenhiser et al. 1989). Because this response can be blocked by L-type Ca^{2+} channel blockers, it appears that activation of dihydropyridine-sensitive voltage-gated Ca^{2+} channels is essential for the elevation of intracellular calcium (Carmines and Navar 1989; Navar et al. 1996). Recent in vivo studies in rat suggest that in the EA angiotensin II-induced vasoconstriction is independent of extracellular calcium influx; instead, the calcium-signaling mechanism involves IP_3-dependent calcium release from intracellular stores (Ruan and Arendshorst 1996).

Inside the glomerulus, mesangial cells have been considered to act in a way similar to the preglomerular smooth muscle cells (Riser et al. 1992). In response to increased tension (resulting from increased stretching by increasing glomerular pressures), mesangial cells were thought to contract, counteracting the expansion of the glomerular tuft. Voltage-dependent L-type Ca^{2+} channel activity has been shown in mesangial cells (McDermott et al. 1993; Yu et al. 1989; Orth et al. 1996). However, signals from the mesangium tuning the efferent resistance have never been seriously considered.

From a structural point of view podocytes would also be an appropriate candidate to perceive tension. Podocytes are regarded as a specific kind of pericyte, which is situated at the outermost circumference of glomerular capillaries and thus at the most appropriate site to sense capillary wall tension. Podocytes in conjunction with the GBM have been considered as the main system for stabilizing capillary diameter in the face of local changes in transmural pressure gradients by changing the tone of their foot process system (Kriz et al. 1994a,b, 1995). Again, however, a specific signal from podocytes acting on the efferent resistance has never been considered. The present study did not address this question and does not provide any evidence for such a mechanism.

Flow. Flow is sensed by endothelial cells as shear stress. The relevant hemodynamic parameters responsible for changes in shear stress are flow velocity and fluid (blood) viscosity (we will leave out the problem of nonlaminar flows). It appears that endothelial cells in general are able to sense shear stress and react to changes in shear stress (Ando and Kamiya 1993), and there are specific data that glomerular endothelial cells are sensitive to shear stress (Ott et al. 1993). Signals from the endothelium are thought to act locally on the mesangium, participating in the function of the mesangium to counteract changes in intraglomerular pressure. However, signals from the intraglomerular endothelium to the EA have never been considered.

From a structural point of view one segment of the glomerular microvasculature, i.e., the EA_{OS}, would appear to be outstanding in its capability to sense shear stress. This vessel segment appears as a structure specifically designed as a *shear stress receptor*. First, compared to all other arteriolar vessel portions of the glomerular microvasculature, this segment contains a conspicuously narrow portion. Since all the blood leaving the glomerulus has to pass through this segment, flow velocity should be highest at this site. Second, this vessel segment is very short, thus, poorly qualified to control efferent resistance actively. Third, this section contains prominent endothelial cells that are densely arranged and that bulge deeply into the vessel lumen. The flasklike shape of their cell bodies appears to indicate that the cells are constantly launched downstream by the blood flow. This appearance becomes especially prominent in the hydronephrotic kidney (Fretschner et al. 1990; Schnabel et al. 1987). Fourth, endothelial cells produce a variety of signal substances in response to changes in shear stress, among them endothelium-derived relaxing factor or nitric oxide (EDRF/NO).

Endothelial cells generally contain an endothelial-specific NO synthase (NOS III) (Förstermann et al. 1991; Moncada et al. 1991). The EA_{OS} – in addition to the endothelial type of NO synthase – has been shown to produce the neuronal type of NO synthase (NOS I) (Bachmann et al. 1995). This is the only vessel segment in the kidney that contains this type of NO synthase. Taken together, this vessel segment has several striking features that strongly suggest that it functions as a shear-stress receptor.

Shear stress is mainly dependent on flow (flow velocity) and viscosity. Any changes in filtration will result in alterations in both parameters. If the filtration fraction decreases, blood velocity at the EA_{OS} will increase and blood viscosity will decrease. Thus, changes in these parameters depend on the actual proportion of filtered fluid, the filtration fraction. Thus, both parameters – viscosity and flow velocity – would be suitable local signals to inform the EA about the efficacy of filtration and to regulate the efferent resistance in response to changes in filtration fraction.

The possible details of such a sensing system need some discussion. First, when there is any filtration, flow velocity in the EA will decrease, resulting in a decrease in shear stress at the EA; blood viscosity, on the other hand, will increase, followed by an increase in shear stress. Thus, it would be necessary to know which of the two possible signals has the major impact on shear stress. Assuming a systemic hematocrit of 50%, a plasma protein concentration of 8%, and a filtration fraction of 20%, flow velocity at the EA will be decreased by about 10%, the hematocrit will rise to about 55.5%, and the plasma protein concentration to about 10%. Recent work in mesenteric vessels has shown that the resistance to flow in microvascular beds is much more sensitive to changes in hematocrit than previously suggested (Pries et al. 1994). Because of these findings blood viscosity has gained more relevance as a determinant of peripheral perfusion and should be considered a major candidate involved in peripheral blood-flow regulation. Thus, there are reasons to assume that blood viscosity is a dominant factor in eliciting changes in shear stress in the EA. Taken together, the EA_{OSi} appears to be an ideal sensor of changes in shear stress based upon changes in blood viscosity brought about by changes in filtration fraction.

The Intraglomerular Segment of the EA as an Effector? On the other hand, the EA_{OSi} does not seem to be a proper effector site to change efferent resistance (R_{EA}) in response to local signals coming from its endothelium. The EA_{OSi} is never encompassed by smooth muscle cells; only if we take the sporadic smooth muscle cells and the mesangial cells together is a complete surrounding layer of contractile cells present. However, the possibility that this layer acts like a smooth muscle layer (tunica media) of a proper arteriole is questionable. Mesangial cells contain much less actin filament bundles than smooth muscle cells. In addition, the filament bundles in mesangial cells are different because of the apparent lack of smooth muscle α-actin (Elger et al. 1993) and the presence of several other contractile proteins (Ishino et al. 1991; Johnson et al.

1991; Kimura et al. 1995). Thus, compared to the smooth muscle layer of a proper arteriole, the contractile forces, which possibly may be created by the mesangial layer around the EA_{OSi}, are probably much less powerful. Moreover, the arrangement and shape of mesangial cells suggest that the effect of a contraction on the capillaries is different from that performed by smooth muscle cells. The latter may completely encircle a vessel, allowing direct constriction. In contrast, mesangial cells generally do not form rings around the glomerular capillaries, but act in conjunction with the GBM to which they are attached. Contraction of mesangial cells will lead to an increased centripetal pull at the GBM, thus indirectly affecting the vessel by increasing the surrounding pressure (Kriz et al. 1990a,b).

In conclusion, the intraglomerular segment of the EA does not appear as a proper effector site. Considering this segment as a specific shear stress receptor, the lack of an active role makes sense. Resistance – fixed within certain boundaries – is less dependent on presettings by an external vasoactive substance that would change its sensitivity. Thus, 'fixed' resistance provides a stable situation to sense changes in shear stress resulting from upstream events.

A Proposed Feedback System Between the Glomerulus and the EA. Taken together, the two segments of the EA appear to play different roles in regulating R_{EA}. The downstream segment, as is well established, represents the effector; the intraglomerular segment, as proposed here, represents a specific sensor site. Because of its narrow lumen and prominent endothelium, the intraglomerular segment appears appropriate for functioning as a shear stress receptor. However, owing to the lack of a complete smooth muscle sheath and its shortness, it is probably not an important player in changing the efferent resistance. Downstream segments of the EA, which directly follow the EA_{OS}, are completely encircled by smooth muscle and appear to be much more capable of actively changinge the vessel diameter. Therefore, a signalling mechanism from the shear stress receptor at the glomerular exit to the downstream segment of the EA would be necessary.

If we consider the problem on a large scale, any signal from inside the glomerulus correlating the efficacy of filtration with the efferent resistance would make sense. Using this broader point of view, there are several candidates: prostaglandins, endothelin, bradykinin, and endothelium-derived hyperpolarizing factor (EDHF). In the present context, we will only discuss possibilities originating from the intraglomerular segment of the EA. First, we will discuss shear stress and NO.

NO is generally considered the most important humoral compound involved in modulation of vascular resistance in response to changes in shear stress. Locally produced in endothelial cells in response to increased shear stress, NO causes relaxation of adjacent smooth muscle cells. Endothelial cells at the EA_{OS} and the downstream segments specifically express neuronal NO synthase (NOS I) in addition to endothelial NO synthase (NOS III) (Bachmann et al. 1995). Since the narrow segment itself is surrounded by mesangial and extraglomerular mesangial (EGM) cells rather than by smooth muscle cells, no local hemodynamically relevant effect of NO produced at this site is obvious. One can speculate that NO at the narrow segment may join other vasoactive substances in defining the contractile tone of mesangial and extraglomerular mesangial cells, but this effect could hardly account for the quick and extensive changes in efferent resistance clearly seen in many in vivo and in vitro preparations (Steinhausen et al. 1990; Steinhausen and Endlich 1995, 1996; Casellas and Moore 1990; Moore and Casellas 1994).

On the other hand, transport of locally produced NO by the blood to more distal effector sites is generally considered unlikely because of its inactivation by hemoglobin (Lancaster 1994; Wood and Garthwaite 1994). The critical question in this context is whether the downstream segments of the EA are still in the effector range of NO produced inside the glomerulus. The downstream segments, which are surrounded by a complete smooth muscle layer, begin within 5–15 μm behind the narrow segment; the entire length of the EA of superficial glomeruli amounts to approximately 100 μm. These are quite short distances, justifying the assumption that NO produced at the narrow segment may still be active when reaching the efferent arteriole proper. Moreover, since specific neuronal NO synthase has been found not only in the narrow segment but also in the downstream segments of the EA (Bachmann et al. 1995), it may be speculated that NO in both sites may join in acting on the entire EA. Moreover, recent work suggests that NO bound in the form of nitrosothiols may be more stable and, especially when bound to cysteine, a very potent vasodilator (Jia et al. 1996); such bound forms of NO may travel with the blood. Studies on the hydronephrotic kidney and the blood-perfused juxtamedullary nephron preparation clearly exhibit reinforcement of an angiotensin II effect in the EA after inhibition of NO synthase by amplifying efferent vasoconstriction (Hoffend et al. 1993; Endlich et al. 1995; Ikenaga et al. 1996a).

Recently, it has been shown that prostaglandins (PG) produced by the glomerulus itself during orthograde perfusion have a profound impact on efferent resistance by decreasing the sensitivity of the downstream EA to angiotensin II and norepinephrine (Arima et al. 1994). The authors speculate about a novel feedback mechanism in which the glomerulus controls its own capillary pressure by releasing PGs and thereby adjusting the resistance of the downstream EA. The mesangial cells were considered as a possible origin for these PGs, but there may, in addition, be other sites of production. PGs would tolerate the short trip to downstream segments of the EA.

Another vasodilative compound is endothelium-derived hyperpolarizing factor (EDHF), which acts via the opening of tetraethylammonium (TEA)-sensitive Ca-activated K^+ channels. Release of this factor from the endothelium is triggered by receptor-dependent agonists such as acetylcholine, bradykinin, histamine or substance P (Feletou and Vanhoutte 1988; Hecker et al. 1994; Campbell et al. 1996). EDHF has been characterized as a cytochrome P450-derived arachidonic acid metabolite (Hecker et al. 1994) and recently identified as an epoxyeicosatrienoic acid (EET) (Campbell et al. 1996). In rat kidney, 11,12 EET is a potent arteriolar dilator (Zou et al. 1994). The vasodilative effect of bradykinin, one major agonist of EDHF, especially on the EA has been demonstrated in a variety of studies (Edwards 1985; Kon et al. 1993; Heller et al. 1994; Endlich et al. 1995).

Gatzek's "Gamma Loop" and Some More Speculations. In several papers, Gatzek (1987, 1993; Gatzek and Wells 1989) has published physical experiments performed with a looped rubber tube, the "gamma loop," serving as a simplified model of the glomerular vascular system. A short rubber tube is formed into a loop by holding it closely together with an elastic cuff (made up of the same kind of rubber tubing) around both branches. One end, representing the afferent branch, is connected to a tank of water that can be raised in height to provide pressure. The efferent branch has a clamp at its end that enables it to be closed or opened in repeatable steps. When pressurized, this system shows spectacular consequences: tightening the outflow clamp stepwise leads to a sudden and quite dramatic increase in the rate of perfusion; thus, the system does not behave according to Ohm's law. The basis of this "gamma loop" phenomenon is obviously the fact that increasing outflow resistance can force open upstream elastic resistance. During the entire maneuver pressure in the loop stays fairly constant.

Remarkable is also the behavior of this system in experiments when returning from higher to lower levels of outflow resistance. The system exhibits hysteresic behavior – probably because the elastic resistance inside the cuff has the tendency to behave like elastic hysteresis. Thus, in this experiment, a high flow rate is maintained at settings of the downstream resistance (the outflow clamp) at which flow was still down when gradually changing outflow resistance in the opposite direction. Thus, the system has a buffer capacity to stabilize flow when the outflow resistance falls.

There is a similarity between the "gamma loop" made up of rubber tubing and the arrangement of the two arterioles at the glomerular hilum: (1) the afferent arteriolar pressure (P_{AA}) may directly influence efferent resistance (R_{EA}) within the glomerular hilum by compressing the lumen of the EA_{OS} and (2) the EA has two types of resistance, one (probably) 'elastic' resistance in the outflow segment (R_{OS}) and a second type of resistance in the distal segment of the EA (R_{DS}) based on the usual smooth muscle cells in the vascular system.

Glomerular filtration critically depends on pressure *and* flow. If we think that P_{AA}, the entrance pressure to the loop, i.e., to glomerular capillaries, is stabilized by preglomerular resistances, the arrangement of two subsequent, qualitatively different types of resistance in the EA would tend – according to Gatzek's experiments – to stabilize pressure and flow in glomerular capillaries within a considerable range of variation in the downstream resistance of the EA. According to Gatzek's model, such an effect would especially be obvious in situations when the downstream resistance decreases.

Whether the glomerular arterioles behave or at least tend to behave in this way is totally unknown. It has been shown (Fretschner et al. 1990; Gabbai et al. 1993; Häberle 1988) that a considerable proportion of resistance in the EA is independent of any smooth muscle tone, suggesting that there are two types of resistance in the EA present, an elastic resistance and a resistance dependent on smooth muscle tone. In the above discussion we suggested that because of the structure of the EA_{OS} the resistance at this site may not change dramatically, as it operates within a narrow range with fixed upper and lower limits.

If we add to Gatzek's hypothesis the shear stress hypothesis, a feedback system between the outflow resistance (R_{OS}) and the downstream resistance (R_{DS}) can be proposed. In a given setting of R_{OS} and a given flow and viscosity, a signal substance (e.g., NO) is produced at a certain rate by the endothelial cells in the EA_{OS}. NO may travel with the blood and reach the downstream segment, decreasing its muscle tone; consequently the downstream resistance R_{DS} will fall. This will lead to a decrease in pressure in the EA, followed possibly by an increase in R_{OS} (because the balance between the pressure in the EA_{OS} and in the AA has shifted in favor of P_{AA} and, thus, to a slightly larger pressure difference across the wall of the EA_{OS}). This, in turn, will increase the shear stress in the EA_{OS}, followed by increasing NO production and closing a positive feedback cycle. However, owing to the structural arrangement of the vascular pole, the cycle will soon reach an upper limit. In contrast to the conditions in Gatzek's rubber tube, it appears very unlikely that the EA_{OS} could ever be totally compressed. We would suggest that the architecture guarantees a lower limit of luminal diameter and, consequently, an upper limit of resistance. Thus, the system would reach stability at an "a priori" given upper limit of R_{OS}.

Now let us consider the opposite situation. Because of some mechanism (increasing circulating antiotensin II, local increase in any other vasoconstrictive substance) the downstream resistance will be increased by increasing the smooth muscle tone at this site (overriding the NO effect). In this case, the pressure in the EA will rise and will possibly lead to some widening of the EA_{OS} and, thus, to a decrease in R_{OS}. This would cause a decrease in shear stress, a decrease in NO secretion and lead to a further increase in R_{DS} by shifting the balance between vasoconstrictive and vasodilative substances towards the vasoconstrictive site. Again, a positive feedback cycle would come into operation. Like the above situation, widening of the EA_{OS} and consequently the decrease in R_{OS} would reach a limit guaranteed by the architecture. The system will stabilize now with opposite settings, minimal R_{OS}, maximal R_{DS}.

In summary, the system would reach stability at the end of both cycles. In one case, R_{OS} would be maximal and R_{DS} minimal; in the other case, the opposite settings would add up to a total efferent resistance. This sum may well be different at both ends, but at both ends it should have a definite maximal value. Seen from this viewpoint, in this system total efferent resistance can never fall to zero and never become exceedingly high. Thus, the system would guarantee that (within certain physiological ranges) neither flow nor pressure can fall to zero or become exceedingly high.

4.3 Integration of Vascular Pole Structures into the Juxtaglomerular Apparatus

The description of the vascular pole as a functional unit overlaps with the concept of the juxtaglomerular apparatus (JGA). The only component of the JGA not included as a component of the vascular pole is the MD. The functional concept of the JGA has developed from the view of a feedback system connecting tubular reabsorption with glomerular filtration based on a signal from the tubule to the glomerulus. The prominent attachment site of the MD to the glomerular vascular pole has dominated research in this field.

The present description emphasizes the vascular pole as a conjoint structure that provides capabilities to regulate glomerular blood flow (and thus filtration) and to protect the high-pressure compartment of the glomerular tuft from expansion. In a first approach these functions may be seen fully independently of any signal from the tubule. The present evaluation of this issue without the aspect "macula densa" permits a more focused discussion of the structural prerequisites of glomerular blood-flow regulation. To add a signal from the MD to this system is not so much a structural question (the attachment of the MD to the pole is structurally organized in a very simple way) as a basically functional question. In contrast, the machinery itself that allows glomerular blood-flow regulation is, first of all, a structural problem. Precise knowledge of the structural organization is a "sine qua non" for recognizing how glomerular blood-flow regulation operates.

4.3.1 What Is the Function of the EGM?

When adding the MD to the system, two aspects receive particular weight: the EGM and the granular cells. Structurally, the EGM and its relationships to the macula densa have been fairly well known for a long time (Barajas et al. 1990; Kriz and Kaissling 1992). The present study adds some more details about the integration of the EGM into the vascular structures at the pole.

Blood flow into the glomerular tuft is locally controlled by two major signals emerging from (1) changes in blood pressure in the AA and (2) changes in tubular fluid composition at the MD. In all likelihood, the site of a possible integration of both signals is the EGM. The EGM cells are located in the center of the vascular pole and would appear to be the most probable candidate for responding to pressure changes. The force necessary to hold the structures of the vascular pole together should be directly proportional to the blood pressure in the glomerulus. This force may be primarily developed by the EGM cells. The sum of expanding forces exerted to the pole (which ultimately are all derived from the blood pressure) should directly translate into tension of EGM cells. Since EGM cells are interconnected among themselves and are connected to all other smooth muscle and smooth muscle-related cells at the vascular pole, direct transmission of signals from the EGM cells to all possible effector cells, including the granular cells, would be possible. In addition, they are the structures most regularly and intimately connected to the MD (Barajas 1970; Barajas et al. 1990; Christensen et al. 1975; Christensen and Bohle 1978; Gorgas 1978; Kriz and Kaissling 1992).

Both signals, changes in blood pressure and changes in tubular fluid composition, are followed by two major responses: first, changes in the AA tone and, second, changes in renin secretion by granular cells (Schnermann and Briggs 1992).

Let us first consider the pertinent features of the MD mechanism. It is widely agreed that the MD functions as a local sensor of tubular fluid composition. Tubular fluid composition is highly dependent on tubular flow. In low-flow states, salt concentration falls; it increases in high-flow states. The change in vascular tone and the change in renin secretion caused by changes in tubular fluid composition at the MD are blocked by agents such as furosemide that interfere with the Na,K,2Cl-cotransport system in MD cells. This has led to the proposal that JGA-mediated control of vascular tone and renin secretion is initiated by concentration-dependent transepithelial transport of NaCl. The mechanism by which transport of NaCl causes changes in the functional state of vascular smooth muscle cells and granular cells is unclear (Schnermann and Briggs 1992; Schricker and Kurtz 1995; Kurtz 1995; Wilcox et al. 1996; Thorup and Persson 1996; Ikenaga et al. 1996b).

Two possibilities are discussed. Either the transport of NaCl itself causes changes in the functional state of EGM cells by changing salt concentration in their surroundings (EGM cells are the primary sensor), or the NaCl transport elicits the generation of a chemical transmitter in the transporting epithelial cells that is released into the basal interstitium, i.e., into the EGM (in this case, the MD cells would be the primary sensor). The recent localization of a Na-K-2Cl cotransporter (BSC2/NKCC1) to the juxtaglomerular portion of the AA and to mesangial cells and EGM cells indicates a new possible mechanism by which Cl^- accumulation in the juxtaglomerular interstitium could be part of a direct signaling mechanism between the MD and the EGM and/or the AA (Kaplan et al. 1996).

The responses in EGM cells to changes in transepithelial NaCl transport are largely unknown (Okuda et al. 1991; Schnermann and Briggs 1992; Matsunaga et al. 1994). It may be suggested that an elevation of intracellular Ca^{2+} concentration, initiated mainly by IP_3-mediated release from intracellular stores, and an increase in Ca^{2+} influx through voltage-dependent Ca^{2+} channels are the underlying mechanisms, as they have been demonstrated to be present in both mesangial and arteriolar cells (and EGM cells may possibly share this property; McDermott et al. 1993; Yu et al. 1989). The extensive coupling between all non-epithelial cells of the JGA by gap junctions provides a pathway through which changes in intracellular Ca^{2+} and membrane potential could extend to vascular smooth muscle cells as well as granular cells (Boll et al. 1975; Pricam et al. 1974; Taugner et al. 1978). Functional evidence for this kind of coupling has not been provided so far.

Renin secretion from granular cells at the vascular pole is pressure-dependent (Kirchheim et al. 1990; Bock et al. 1992; Kurtz 1995). When perfusion pressure in the AA decreases, renin release will rise, beginning sharply at a certain pressure level. It is widely believed that the granular cells themselves are the sensors being stretched in proportion to the transmural pressure gradient. We have discussed above that, although such a mechanism appears possible, the irregular and variable arrangement of granular cells in the wall of the AA is not ideal for such a mechanism. It is hard to imagine that the granular cells would all be affected by a pressure increase to a similar extent, resulting in sensing the same degree of tension. Therefore, the most reasonable structure to sense changes in hydrostatic pressure at the glomerular hilum is the EGM cells. Their degree of tension should be strictly correlated with the pressure in the AA.

The integration of pressure-dependent signals with MD signals would seem to be possible by summing up intracellular Ca^{2+} concentrations. Mesangial cells and vascular smooth muscle cells are known to contain mechanosensitive voltage-dependent and mechanosensitive voltage-independent Ca^{2+} channels (EGM cells may readily be expected to have them as well (Chen et al. 1994; Krause and Nobiling 1994; Paßreiter and Nobiling 1993). The various types of channels and other possible activation mechanisms would cooperate in EGM cells to integrate signals from tubular fluid composition and AA pressure. Thus, the EGM would be the necessary integrator.

4.4 Fluid Leakage Through the Glomerular Stalk

It is widely believed that any surplus of glomerular interstitial fluid contained in the mesangium might drain via the glomerular stalk into the cortical interstitium (Makino et al. 1994; Michael et al. 1980). Evidence for such a draining route comes from studies in which marker substances were followed along their hypothetical pathway from the mesangium, through the stalk, through the EGM, finally into the cortical interstitium; however, the evidence in these studies is very weak.

The present findings do not support such a drainage route or, at least, would not give much relevance to it. The mesangial space inside the glomerular stalk is very narrow, mesangial channels (which are often mentioned in the literature; [Michael et al. 1980; Raij and Keane 1985]) do not exist; and free channels occur when the matrix is dissolved (Makino et al. 1994). If we think of the enormous blood flow through the glomerular tuft and the fact that the capillary-mesangial interface is very leaky, fluid entry into the mesangium should be enormous. This fluid must have other ways of leaving the mesangium than through the stalk. We would suggest that any fluid entry into the mesangium is constantly balanced by an equal exit of fluid across the paramesangial barrier into Bowman's space (Lemley et al. 1992).

In this and previous studies (Rosivall et al. 1991) it has been shown that the endothelium changes into the fenestrated type very early along the AA. These porous parts regularly face the EGM. Thus, fluid leakage from the vessel into the surroundings should occur not only in glomerular capillaries (through the endothelio-mesangial interface into the intraglomerular mesangium) but, in addition, directly from the AA into the EGM. Thus, most of the fluid throughput of the EGM is probably derived directly from the AA and from the MD. Compared to these sources, any input via the glomerular stalk would seem to be extremely small at best.

5 Summary

In the present study we provide a detailed structural analysis of the vascular pole of superficial and midcortical glomeruli of the rat kidney. A description of the juxtaglomerular portions of the afferent and efferent arterioles, the extraglomerular mesangium and the glomerular stalk is included. The specific structural elaboration of the epithelial transition from the podocytes to the parietal epithelium is emphasized, with particular attention to the arrangement of the cytoskeleton and its connections to extracellular matrix elements. The branching patterns of the afferent and efferent arterioles are quite different. Immediately at the glomerular entrance, the afferent arteriole divides into its primary branches. In contrast, the efferent arteriole has a specific outflow segment (consisting of an intraglomerular portion and a portion associated with the extraglomerular mesangium) established by the confluence of capillary tributaries deep inside the glomerular tuft. Just at the transition from inside to outside, this segment includes a prominent narrow portion with conspicuous endothelial cells bulging into the vessel lumen. The extraglomerular mesangium has been found to represent a solid block of cells and matrix filling the space between the macula densa and both arterioles and extending into the entrance funnel. Peripherally located extraglomerular mesangial cells attach to the outer aspect of the parietal basement membrane. As a whole, the extraglomerular mesangium occludes the glomerular tuft.

The results appear relevant with respect to four major aspects: (1) a support function counteracting the expansile forces resulting from the high intraglomerular pressures, (2) a direct functional influence of the afferent on the efferent arteriole, resulting from their narrow assemblage at the glomerular entrance, (3) a specific shear stress receptor function of the intraglomerular segment of the efferent arteriole, and (4) fluid leakage from the glomerular tuft through the stalk and the extraglomerular mesangium into the cortical interstitium.

1. The glomerulus is a high-pressure compartment; expansile forces continuously tend to expand glomerular capillaries, the glomerular stalk, and the glomerular entrance. Counteracting centripetal forces at the vascular pole appear to be developed as circular forces by the cytoskeleton of podocytes and parietal cells surrounding the glomerular entrance and as interconnecting forces between both arterioles and between opposing walls of the glomerular entrance, as well as of the glomerular stalk. These interconnecting forces are developed by the extraglomerular mesangium which – as a whole – forms a spiderlike closure device holding the glomerular entrance together. In addition, the extraglomerular mesangium develops occluding forces, allowing a gradual pressure drop between the glomerular stalk and the macula densa.

2. At the glomerular entrance, the outflow segment of the efferent arteriole is narrowly associated with the bifurcation of the afferent arteriole. Both are enclosed together in a common compartment surrounded by the glomerular basement membrane; there is no pressure barrier individually encompassing each vessel. Therefore, it may readily be suggested that the hydrostatic pressure of the afferent arteriole acts on the efferent arteriole. As a consequence, the luminal width of the efferent arteriole at this site, i.e., its resistance, may be directly modified by the pressure in the afferent arteriole.
3. The efferent arteriole at the transition of the intraglomerular segment to the segment that passes through the extraglomerular mesangium has a conspicuously narrow portion with endothelial cells protruding into the vessel lumen. In addition, this segment is prominent by the expression of the neuronal type of nitric oxide synthase. We therefore propose that this segment acts as a specific shear stress receptor. The possible relevance of a shear stress receptor at this site would be to measure the gain of glomerular filtration. Dependent on the filtration rate and fraction, viscosity and velocity of the effluent blood in the outflow segment will change, resulting in changes in shear stress. Subsequent responses of downstream segments leading to changes in luminal diameter would establish a direct feedback between the filtration gain and the efferent resistance.
4. It is widely believed that any surplus of glomerular interstitial fluid contained in the mesangial compartment of the glomerular tuft drains via the glomerular stalk into the cortical interstitium. The present study does not support this assumption. The mesangial space within the glomerular stalk is extremely narrow; mesangial channels do not exist. We would suggest that any fluid entry into the mesangium is constantly balanced by the exit of an equal amount of fluid across the paramesangial barrier into Bowman's space. The fluid passing the extraglomerular mesangium probably enters mainly from the afferent arteriole and the macula densa. Compared to these sources, any input via the glomerular stalk would appear to be extremely small at best.

The present study emphasizes the vascular pole as a conjoint structure that provides capabilities to regulate glomerular blood flow and filtration and to protect the high-pressure compartment of the glomerular tuft from expansion. We have refrained from including a description of the topographical relationships to the macula densa; this topic has been repeatedly reviewed. The extraglomerular mesangium is the only structure of the vascular pole that consistently contacts the macula densa. Moreover, the extraglomerular mesangium is the central structure interconnecting all other components at the vascular pole. Thus, the extraglomerular mesangium would be qualified to integrate hemodynamic signals coming from both glomerular arterioles and signals about urine composition from the macula densa.

References

Allison MEM, Wilson CB, Gottschalk CW, Davis SK (1974) Pathophysiology of experimental glomerulonephritis in rats. J Clin Invest 53:1402–1423

Ando J, Kamiya A (1993) Blood flow and vascular endothelial cell function. Front Med Biol Eng 5:245–264

Arima S, Ren Y, Juncos LA, Carretero OA, Ito S (1994) Glomerular prostaglandins modulate vascular reactivity of the downstream efferent arterioles. Kidney Int 45:650–658

Aukland K (1989) Myogenic mechanisms in the kidney. J Hypertens 7:S71–S76

Bachmann S, Kriz W, Kuhn C, Franke WW (1983) Differentiation of cell types of the mammalian kidney by immunofluorescence microscopy using antibodies to intermediate filament proteins and desmoplakins. Histochemistry 77:365–394

Bachmann S, Bosse HM, Mundel P (1995) Topography of nitric oxide synthesis by localizing constitutive NO synthases in mammalian kidney. Am J Physiol 268:F885–F898

Barajas L (1970) The ultrastructure of the juxtaglomerular apparatus as disclosed by three-dimensional reconstrucion from serial sections: the anatomical relationship between the tubular and vascular components. J Ultrastruct Res 33:116–147

Barajas L, Latta H (1963) A three-dimensional study of the juxtaglomerular apparatus in the rat. Light and electron microscopic observations. Lab Invest 12:257–269

Barajas L, Salido EC, Powers KV (1990) The juxtaglomerular apparatus. A morphologic perspective. In: Laragh JH, Brenner BM (eds) Hypertension: pathophysiology, diagnosis and management. Raven, New York, pp 1053–1065

Barajas L, Liu L, Tucker M (1994) Localization of connexin43 in rat kidney. Kidney Int 46:621–626

Beeuwkes R (1971) Efferent vascular patterns and early vascular-tubular relations in the dog kidney. Am J Physiol 221:1361–1374

Blanc-Brunat N, Mutin M, Peyrol S (1989) Immunohistochemical localization of type IV collagen fibronectin and laminin in the juxtaglomerular apparatus of the rat kidney. Cell Mol Biol 35:469–484

Bock H, Hermle M, Brunner FP, Thiel G (1992) Pressure dependent modulation of renin release in isolated perfused glomeruli. Kidney Int 41:275–280

Boll HU, Forssmann WG, Taugner R (1975) Studies on the juxtaglomerular apparatus. IV. Freeze-fracturing of membrane surfaces. Cell Tissue Res 161:459–469

Bonsib SM (1986) The macula densa tubular basement membrane: a unique plaque of basement membrane specialization. J Ultrastruct Mol Struct Res 97:103–108

Breuss JM, Gillett N, Lu L, Sheppard D, Pytela R (1993) Restricted distribution of integrin β6 mRNA in primate epithelial tissue. J Histochem Cytochem 41:1521–1527

Briggs JP, Schnermann JB (1987) The tubuloglomerular feedback mechanism: functional and biochemical aspects. Annu Rev Physiol 49:251–273

Burkholder PM, Hyman LR, Barber TA (1973) Extracellular clusters of spherical microparticles in glomeruli in human renal glomerular diseases. Lab Invest 28:415–425

Burridge K, Fath K, Kelly T, Nuckolls G, Turner C (1988) Focal adhesion: transmembrane junctions between the extracellular matrix and the cytoskeleton. Annu Rev Cell Biol 4:487–525

Busk M, Pytela R, Sheppard D (1992) Characterization of the integrin αvβ6 as a fibronectin-binding protein. J Biol Chem 267:5790–5796

Butkowski RJ, Wieslander J, Kleppel M, Michael AF, Fish AJ (1989) Basement membrane collagen in the kidney: Regional localization of novel chains related to collagen IV. Kidney Int 35:1195–1202

Campbell WB, Gebremedhin D, Pratt PF, Harder DR (1996) Identification of epoxyeicosatrienoic acids as endothelium-derived hyperpolarizing factors. Circ Res 78:415–423
Carlemalm E, Garavito RM, Villinger W (1982) Resin development for electron microscopy and an analysis of embedding at low temperature. J Microsc 126:123–143
Carmines PK, Navar LG (1989) Disparate effects of Ca channel blockade on afferent and efferent arteriolar responses to ANG II. Am J Physiol 256:F1015–F1020
Casellas D, Moore LC (1990) Autoregulation and tubuloglomerular feedback in juxtamedullary glomerular arterioles. Am J Physiol 258:F660–F669
Chen V, Guber HA, Palant CE (1994) Mechanosensitive single channel calcium currents in rat mesangial cells. Biochem Biophys Res Commun 203:773–779
Christensen JA, Bohle A (1978) The juxtaglomerular apparatus in the normal rat kidney. Virchows Arch [A] Pathol Anat Histol 379:143–150
Christensen JA, Meyer D, Bohle A (1975) The structure of the human juxtaglomerular apparatus. A morphometric, light microscopic study on serial sections. Virchows Arch [A] Pathol Anat Histol 367:83–92
Couchman JR, Beavan LA, McCarthy KJ (1994) Glomerular matrix: synthesis, turnover and role in mesangial expansion. Kidney Int 45:328–335
Cybulsky AV, Carbonetto S, Huang Q, McTavish AJ, Cyr M-D (1992) Adhesion of rat glomerular epithelial cells to extracellular matrices: Role of β1 integrins. Kidney Int 42:1099–1106
Davis JM (1991) Role of the efferent arteriole in tubuloglomerular feedback. Kidney Int 39 [Suppl 32]:S71–S73
Dieterich HJ (1978) Die Struktur der Blutgefäße in der Rattenniere. In: Bargmann W, Doerr W (eds) Normale und pathologische Anatomie, vol 35. Thieme, Stuttgart, pp 1–127
Downie I, Gardiner DS, Downie TT, Gibson IW, Kenyon C, More IAR, Lindop GBM (1992) Non-granulated peripolar cells exist in the rat glomerulus. Cell Tissue Res 268:567–570
Drumond MC, Deen WM (1994) Structural determinants of glomerular hydraulic permeability. Am J Physiol 266:F1–F12
Edwards RM (1983) Segmental effects of norepinephrine and angiotensin II on isolated renal microvessels. Am J Physiol 244:F526–F534
Edwards RM (1985) Response of isolated renal arterioles to acetycholine, dopamine, and bradykinin. Am J Physiol 248:F183–F189
Elger M, Sakai T, Winkler D, Kriz W (1991) Structure of the outflow segment of the efferent arteriole in rat superficial glomeruli. Contrib Nephrol 95:22–33
Elger M, Drenckhahn D, Nobiling R, Mundel P, Kriz W (1993) Cultured rat mesangial cells contain smooth muscle α-actin not found in vivo. Am J Pathol 142:497–509
Endlich K, Forssmann WG, Steinhausen M (1995) Effects of urodilatin in the rat kidney: comparison with ANF and interaction with vasoactive substances. Kidney Int 47:1558–1568
Endlich K, Hoffend J, Steinhausen M (1996) Localization of endothelin ETA and ETB receptor-mediated constriction in the renal microcirculation of rats. J Physiol (Lond) 491:211–218
Faarup P (1965) On the morphology of the juxtaglomerular apparatus. Acta Anat 60:20–38
Feletou M, Vanhoutte PM (1988) Endothelium-dependent hyperpolarization of canine coronary smooth muscle. Br J Pharmacol 93:515–524
Fischer E, Schnermann JB, Briggs JP, Kriz W, Ronco PM, Bachmann S (1995) Ontogeny of NO synthase and renin in juxtaglomerular apparatus of rat kidneys. Am J Physiol 268:F1164–F1176
Fleming JT, Parekh N, Steinhausen M (1987) Calcium antagonists preferentially dilate preglomerular vessels of hydronephrotic kidney. Am J Physiol 253:F1157–F1163
Förstermann U, Schmidt HH, Pollock JS, Sheng H, Mitchell JA, Warner TD, Nakane M, Murad F (1991) Isoforms of nitric oxide synthase. Characterization and purification from different cell types. Biochem Pharmacol 42:1849–1857
Frank M, Kriz W (1982) Scanning electron microscopy studies of the vascular pole of the rat glomerulus. Anat Rec 204:149–152
Fretschner M, Endlich K, Fester C, Parekh N, Steinhausen M (1990) A narrow segment of the efferent arteriole controls efferent resistance in the hydronephrotic rat kidney. Kidney Int 37:1227–1239
Gabbai FB, Peterson OW, Wilson CB, Blantz RC (1993) Glomerular hemodynamics in the cell free (IPK) and erythrocytes perfused (IEPK) isolated perfused kidney. J Am Soc Nephrol 4:578

Gall JAM, Alcorn D, Butkus A, Coghlan JP, Ryan GB (1986) Distribution of glomerular peripolar cells in different mammalian species. Cell Tissue Res 244:203–208
Gatzek WJ (1987) The gamma loop, a model of glomerular hemodynamics. Med Hypotheses 24:137–141
Gatzek WJ (1993) The elastic constraint effect: a possible factor in renal perfusion. Med Hypotheses 41:215–216
Gatzek WJ, Wells SC (1989) The gamma loop: a model of glomerular hemodynamics part II. Med Hypotheses 29:29–31
Gibson IW, Downie I, Downie TT, Han SW, More IAR, Lindop GBM (1992) The parietal podocyte: A study of the vascular pole of the human glomerulus. Kidney Int 41:211–214
Gibson IW, Downie TT, More IAR, Lindop GBM (1996) Atubular glomruli and glomerular cysts – a possible pathway for nephron loss in the human kidney. J Pathol 179:421–426
Gorgas K (1978) Structure and innervation of the juxtaglomerular apparatus of the rat. Adv Anat Embryol Cell Biol 54:5–84
Häberle DA (1988) Hemodynamic interactions between intrinsic blood flow control mechanisms in the rat kidney. Renal Physiol Biochem 11:289–315
Harder DR, Gilbert R, Lombard JH (1987) Vascular muscle cell depolarization and activation in renal arteries on elevation of transmural pressure. Am J Physiol 253:F778–F781
Hayashi K, Epstein M, Loutzenhiser R (1989) Pressure-induced vasoconstriction of renal microvessels in normotensive and hypertensive rats. Studies in the isolated perfused hydronephrotic kidney. Circ Res 65:1475–1484
Hecker M, Bara AT, Bauersachs J, Busse R (1994) Characterization of endohelium-derived hyperpolarizing factor as a cytochrome P450-derived arachidonic acid metabolite in mammals. J Physiol (Lond) 481:407–414
Heller J, Kramer HJ, Horácek V (1994) The effect of kinin and prostaglandin inhibitors on the renal response to angiotensin-converting enzyme inhibition: a micropuncture study in the dog. Pfluegers Arch 427:219–224
Hitomi K, Murakami T, Kaneshige T (1987) Human kidney glomeruli, with special reference to those in the aged person: scanning electron microscopy study of microvascular corrosion casts. Acta Med Okayama 41:165–171
Hoffend J, Cavarape A, Endlich K, Steinhausen M (1993) Influence of endothelium-derived relaxing factor on renal microvessels and pressure-dependent vasodilation. Am J Physiol 265:F285–F292
Hollywell CA, Jaworowski A, Thumwood C, Alcorn D, Ryan GB (1992) Immunohistochemical localization of transthyretin in glomerular peripolar cells of newborn sheep. Cell Tissue Res 267:193–197
Ikenaga H, Fallet RW, Carmines PK (1996a) Basal nitric oxide production curtails arteriolar vasoconstrictor responses to ANG II in rat kidney. Am J Physiol 271:F365–F373
Ikenaga H, Fallet RW, Carmines PK (1996b) Contribution of tubuloglomerular feedback to renal arteriolar angiotensin II responsiveness. Kidney Int 49:34–39
Ishino T, Kobayashi R, Wakui H, Fukushima Y, Nakamoto Y, Miura AB (1991) Biochemical characterization of contractile proteins of rat cultured mesangial cells. Kidney Int 39:1118–1124
Jia L, Bonaventura C, Bonaventura J, Stamler JS (1996) S-nitrosohaemoglobin: a dynamic activity of blood involved in vascular control. Nature 380:221–226
Johnson RJ, Iida H, Alpers CE, Majesky MW, Schwartz SM, Pritzl P, Gordon K, Gown AM (1991) Expression of smooth muscle cell phenotype by rat mesangial cells in immune complex nephritis. J Clin Invest 87:847–858
Johnston WH, Latta H, Osvaldo L (1973) Variations in glomerular ultrastructure in rat kidneys fixed by perfusion. J Ultrastruct Res 45:149–167
Kaissling B, Kriz W (1982) Variability of intercellular spaces between macula densa cells: a transmission electron microscopic study in rabbits and rats. Kidney Int 22 [Suppl 12]:S9–S17
Kaplan MR, Plotkin MD, Brown D, Herbert SC, Delpire E (1996) Expression of the mouse Na-K-2Cl cotransporter, mBSC2, in the terminal inner medullary collecting duct, the glomerular and extraglomerular mesangium, and the glomerular afferent arteriole. J Clin Invest 98:723–730
Kikuta A, Murakami T (1989) Three-dimensional vascular architecture of the malpighi's gomerular capillary beds as studied by vascular corrosion casting-sem method. In: Motta PM (ed) Cells and

tissues: a three-dimensional approach by modern techniques in microscopy. Liss, New York, pp 181–188

Kimura K, Hirata Y, Nanba S, Tojo A, Matsuoka H, Sugimoto T (1990) Effects of atrial natriuretic peptide on renal arterioles: morphometric analysis using microvascular casts. Am J Physiol 259:F936–F944

Kimura K, Nagai R, Sakai T, Aikawa M, Kuro-o M, Kobayashi N, Shirato I, Inagami T, Oshi M, Suzuki N, Oba S, Mise N, Tojo A, Hirata Y, Goto A, Yazaki Y, Omata M (1995) Diversity and variability of smooth muscle phenotypes of renal arterioles as revealed by myosin isoform expression. Kidney Int 48:372–382

Kirchheim HR, Ehmke H, Persson P (1990) Role of blood pressure in the control of renin release. Acta Physiol Scand 139 [Suppl 591]:40–47

Kon V, Fogo A, Ichikawa I (1993) Bradykinin causes selective efferent arteriolar dilation during angiotensin I converting enzyme inhibition. Kidney Int 44:545–550

Krause M, Nobiling R (1994) Myogen reaction and intracellular free calcium in kidney vessels. Pfluegers Arch 426 [Suppl 6:]R107 (abstract)

Kriz W, Bachmann S (1985) Pre- and postglomerular arterioles of the kidney. J Cardiovasc Pharmacol 7 [Suppl 3]:S24–S30

Kriz W, Dieterich HJ (1970) The supplying and draining vessels of the renal medulla in mammals. In: Proceedings of the 4th international congress of nephrology. Karger, Basel, pp 138–144.

Kriz W, Kaissling B (1992) Structural organization of the mammalian kidney. In: Seldin DW, Giebisch G (eds) The kidney: physiology and pathophysiology. Raven, New York, pp 707–777

Kriz W, Sakai T, Hosser H (1987) Morphological aspects of glomerular function. In: Davison AM (ed) Nephrology, vol 1. Proceedings of the Xth international congress on nephrology, London, 1987. Bailliere Tindall, London, pp 3–23

Kriz W, Elger M, Lemley KV, Sakai T (1990a) Structure of the glomerular mesangium: a biomechanical interpretation. Kidney Int 38 [Suppl 30]:S2–S9

Kriz W, Elger M, Lemley KV, Sakai T (1990b) Mesangial cell – glomerular basement membrane connections counteract glomerular capillary and mesangium expansion. Am J Nephrol 10 [Suppl 1]:4–13

Kriz W, Hackenthal E, Nobiling R, Sakai T, Elger M (1994a) A role for podocytes to counteract capillary wall distension. Kidney Int 45:369–376

Kriz W, Mundel P, Elger M (1994b) The contractile apparatus of podocytes is arranged to counteract GBM expansion. Contr Nephrol 107:1–9

Kriz W, Elger M, Mundel P, Lemley KV (1995) Structure-stabilizing forces in the glomerular tuft. J Am Soc Nephrol 5:1731–1739

Kurtz A (1995) Renin and hypertension. Nephrol Dial Transplant 10:1521–1523

Lancaster JR (1994) Simulation of the diffusion and reaction of endogenously produced nitric oxide. Proc Natl Acad Sci USA 91:8137–8141

Lebrun F, Morel F, Vassent G, Marchetti J (1992) Cholenergic effects on intracellular free calcium concentration in renal corpuscle: role of parietal sheet. Am J Physiol 262:F248–F255

Lemley KV, Elger M, Koeppen-Hagemann I, Kretzler M, Nagata M, Sakai T, Uiker S, Kriz W (1992) The glomerular mesangium: capillary support function and its failure under experimental conditions. Clin Invest 70:843–856

Lethias C, Aubert-Foucher E, Dublet B, Eichenberger D, Font B, Goldschmidt D, Labourdette L, Mazzorana M, van der Rest M (1994) Structure, molecular assembly and tissue distribution of facit collagen molecules. Contr Nephrol 107:57–63

Loutzenhiser R, Hayashi K, Epstein M (1989) Divergent effects of KCl-induced depolarization on afferent and efferent arterioles. Am J Physiol 257:F546–F564

Luff SE, Hengstberger SG, McLachlan EM, Anderson WP (1992) Distribution of sympathetic neuroeffector junctions in the juxtaglomerular region of the rabbit kidney. J Auton Nerv Syst 40:239–253

Makino H, Hironaka K, Shikata K, Nagake Y, Kumagai I, Kashihara N, Ota Z (1994) Mesangial matrices act as mesangial channels to the juxtaglomerular zone. Nephron 66:181–188

Matsunaga H, Yamashita N, Okuda T, Kurokawa K (1994) Mesangial cell ion transport and tubuloglomerular feedback. Curr Opin Nephrol Hypertens 3:518–522

Mbassa GK (1991) Characteristics of peripolar cell granules of antelopes and goats. Vet Rec 129:289-291
Mbassa GK, Elger M, Kriz W (1988) The ultrastructural organization of the basement membrane of Bowman's capsule in the rat renal corpuscle. Cell Tissue Res 253:151–163
McDermott GF, Hurst RD, Whiteside CI (1993) Isolated rat glomerular cells demonstrate L-type Ca2+ -channel activity. Cell Calcium 14:387–396
Michael AF, Keane WF, Raij L, Vernier RL, Mauer SM (1980) The glomerular mesangium. Kidney Int 17:141–154
Moncada S, Palmer RMJ, Higgs EA (1991) Nitric oxide: physiology, pathophysiology, and pharmacology. Pharmacol Rev 43:109–142
Moore BJ, Homes KR, Xu LX (1992) Vascular anatomy of the pig kidney glomerulus: a qualitative study of corrosion casts. Scanning Microsc 6:887–898
Moore LC, Casellas D (1990) Tubuloglomerular feedback dependence of autoregulation in rat juxtamedullary afferent arterioles. Kidney Int 37:1402–1408
Murakami T (1972) Vascular arrangement of the rat renal glomerulus. A scanning electron microscope study of corrosion casts. Arch Histol Jpn 34:87–107
Navar LG, Inscho EW, Majid DSA, Imig JD, Harrison-Bernard LM, Mitchell KD (1996) Paracrine regulation of the renal microcirculation. Physiol Rev 76:425–536
Okuda T, Naruse M, Kurokawa K (1991) Mesangial cell function and chloride ions. Kidney Int 39 [Suppl 32]:S65–S67
Olsen S, Bohman SO, Petersen VP (1974) Ultrastructure of the glomerular basement membrane in long term renal allografts with transplant glomerular disease. Lab Invest 30:176
Orth SR, Nobiling R, Bonisch S, Ritz E (1996) Inhibitory effect of calcium channel blockers on human mesangial cell growth: evidence for actions independent of L-type Ca2+ channels. Kidney Int 49:868–879
Ott MJ, Olson JL, Ballermann BJ (1993) Phenotypic differences between glomerular capillary (GE) and aortic (AE) endothelial cells in vitro. J Am Soc Nephrol 4:564 (abstract)
Paßreiter M, Nobiling R (1993) Changes in intracellular calcium after mechanical stimulation of glomerular mesangial cells. Pfluegers Arch 422 [Suppl 1:]R479 (abstract)
Pease DC (1968) Myoid features of renal corpuscles and tubules. J Ultrastruct Res 23:304–320
Petermann A, Fees H, Grenz H, Goodman SL, Sterzel RB (1993) Polymerase chain reaction and focal contact formation indicate integrin expression in mesangial cells. Kidney Int 44:997–1005
Pricam C, Humbert F, Perrelet A, Orci L (1974) Gap junctions in mesangial and lacis cells. J Cell Biol 63:349–354
Pries AR, Secomb TW, Geβner T, Sperandio MB, Gross JF, Gaehtgens P (1994) Resistance of blood flow in microvessels in vivo. Circ Res 75:904–915
Raij L, Keane WF (1985) Glomerular mesangium: its funtion and relationship to angiotensin II. Am J Med 79:24–30
Ren YL, Garvin JJ, Carretero OA (1996) Role of efferent arteriole in tubuloglomerular feedback. J Am Soc Nephrol 7:1587 (abstract)
Riser BL, Cortes P, Zhao X, Bernstein J, Dumler F, Narins RG (1992) Intraglomerular pressure and mesangial stretching stimulate extracellular matrix formation in the rat. J Clin Invest 90:1932–1943
Rosivall L, Rázaga Z, Ormos J (1991) Morphological characterization of human juxtaglomerular apparatus. Kidney Int 39 [Suppl 32]:S9–S12
Ruan X, Arendshorst WJ (1996) Calcium entry and mobilization signaling pathways in ANG II-induced renal vasoconstriction in vivo. Am J Physiol 270 (39):F398–F405
Ryan GB, Coghlan JP, Scoggins BA (1979) The granulated peripolar epithelial cell: a potential secretory component of the renal juxtaglomerular complex. Nature 277:655–656
Sakai T, Kriz W (1987) The structural relationship between mesangial cells and basement membrane of the renal glomerulus. Anat Embryol (Berl) 176:373–386
Sanchez-Ferrer CF, Roman RJ, Harder DR (1989) Pressure-dependent contraction of rat juxtamedullary afferent arterioles. Circ Res 64:790–798
Saxén L (1987) Organogenesis of the kidney. University Press, Cambridge
Schnabel E, Kriz W, Steinhausen M (1987) Outflow segment of the efferent arteriole of the rat glomerulus investigated by in vivo and electron microscopy. Renal Physiol 10:318–326

Schnermann JB, Briggs JP (1992) Function of the juxtaglomerular apparatus: control of glomerular hemodynamics and renin secretion. In: Seldin DW, Giebisch G (eds) The kidney: physiology and pathophysiology. Raven, New York, pp 1249–1290
Schricker K, Kurtz A (1995) Role of membrane-permeable ions in renin secretion by renal juxtaglomerular cells. Am J Physiol 269:F64–F69
Shirato I, Sakai T, Kimura K, Tomino Y, Kriz W (1996) Cytoskeletal changes in podocytes associated with foot process effacement in Masugi nephritis. Am J Pathol 148:1283–1296
Snoei H (1987) Das Glomerulum-Kapillar-Netz der Rattenniere. Netzwerkrekonstruktion und Morphometrie mit Hilfe intravitalmikroskopischer Strukturanalyse. Dissertation, University of Heidelberg
Spanidis A, Wunsch H, Kaissling B, Kriz W (1982) Three-dimensional shape of a Goormaghtigh cell and its contact with a granular cell in the rabbit kidney. Anat Embryol (Berl) 165:239–252
Spinelli FR (1974) Structure and development of the renal glomerulus as revealed by scanning electron microscopy. Int Rev Cytol 39:345–381
Spinelli FR, Wirz H, Brücher Ch, Pehling G (1972) Non-existence of shunts between afferent and efferent arterioles of juxtamedullary glomeruli in dog and rat kidneys. Nephron 9:123–128 (abstract)
Steinhausen M, Endlich K (1995) Control of renal microcirculation. Nephrol Dial Transplant 10:1559–1564
Steinhausen M, Endlich K (1996) Controversies on glomerular filtration from Ludwig to the present. Pfluegers Arch 432:R73–R81
Steinhausen M, Snoei H, Parekh N, Baker R, Johnson PC (1983) Hydronephrosis: a new method to visualize vas afferens, efferens, and glomerular network. Kidney Int 23:794–806
Steinhausen M, Blum M, Fleming JT, Holz FG, Parekh N, Wiegman DL, Dussel R (1989) Visualization of renal autoregulation in the split hydronephrotic kidney of rats. Kidney Int 35:1151–1160
Steinhausen M, Endlich K, Wiegman D (1990) Glomerular blood flow. Kidney Int 38:769–784
Takenaka T, Harrison-Bernard LM, Inscho EW, Carmines PK, Navar LG (1994) Autoregulation of afferent arteriolar blood flow in juxtamedullary nephrons. Am J Physiol 267:F879–F887
Taugner R, Hackenthal E (1989) The juxtaglomerular apparatus. Springer, Berlin Heidelberg New York
Taugner R, Schiller A, Kaissling B, Kriz W (1978) Gap junctional coupling between the JGA and the glomerular tuft. Cell Tissue Res 186:279–285
Thorup C, Persson AEG (1996) Macula densa derived nitric oxide in regulation of glomerular capillary pressure. Kidney Int 49:430–436
Trahair JF, Ryan GB (1989) Co-localization of neuron-specific enolase-like and kallikrein-like immunoreactivity in ductal and tubular epithelium of sheep salivary gland and kidney. J Histochem Cytochem 37:309–314
Unehira M (1981) Blood vascular architecture of the rhesus monkey kidney glomerulus. A scanning electron microscope study of corrosion casts. Okayama Igakkai Zasshi 93:951–961
Waugh D, Prentice RSA, Yadav D (1967) The structure of the proximal tubule: a morphological study of basement membrane cristae and their relationship in the renal tubule of the rat. Am J Anat 121:775–786
Wilcox CS, Deng X, Welch WJ (1996) Roles of neuronal nitric oxide synthase and macula densa reabsorption in enhanced renal nitric oxide generation during salt loading in rats. J Am Soc Nephrol 7:1591 (abstract)
Wilson SK (1986) The effects of angiotensin II and norepinephrine on afferent arterioles in the rat. Kidney Int 30:895–905
Winkler D, Elger M, Sakai T, Kriz W (1991) Branching and confluence pattern of glomerular arterioles in the rat. Kidney Int 39 [Suppl 32]:S2–S8
Wood J, Garthwaite J (1994) Models of the diffusional spread of nitric oxide: implications for neural nitric oxide signalling and its pharmacological properties. Neuropharmacology 33(11):1235–1244
Woods A, Couchman JR (1992) Protein kinase C involvement in focal adhesion formation. J Cell Sci 101:277–290
Yu YM, Lermioglu F, Hassid A (1989) Modulation of Ca by agents affecting voltage-sensitive Ca channels in mesangial cells. Am J Physiol 257:F1094–F1099

Zou AP, Fleming JT, Falck JR, Jacobs ER, Gebremedhin D, Harder DR, Roman RJ (1994) Regional and stereospecific effects of 11,12-epoxyeicosatrienoic acids on vascular tone and K+ channel activity in renal arterioles. Hypertension 24:382 (abstract)

Subject Index

The bold page numbers indicate the figures

Reviews and critical articles covering the entire field of normal anatomy (cytology, histology, cyto- and histochemistry, electron microscopy, macroscopy, experimental morphology and embryology and comparative anatomy) are published in Advances in Anatomy, Embryology and Cell Biology. Papers dealing with anthropology and clinical morphology that aim to encourage cooperation between anatomy and related disciplines will also be accepted. Papers are normally commissioned. Original papers and communications may be submitted and will be considered for publication provided they meet the requirements of a review article and thus fit into the scope of "Advances". English language is preferred, but in exceptional cases French or German papers will be accepted.

It is a fundamental condition that submitted manuscripts have not been and will not simultaneously be submitted or published elsewhere. With the acceptance of a manuscript for publication, the publisher acquires full and exclusive copyright for all languages and countries.

Twenty-five copies of each paper are supplied free of charge.

Manuscripts should be addressed to

Prof. Dr. F. **BECK,** Howard Florey Institute, University of Melbourne, Parkville, 3000 Melbourne, Victoria, Australia

Dr. D. **BROWN**, PhD, Massachusetts General Hospital, The Renat Unit, 149, 13th Street, 8th Floor, Charlestown, Massachusetts 02129, USA

Prof. Dr. B. **CHRIST**, Anatomisches Institut der Universität Freiburg, Abteilung Anatomie II, Albertstr. 17, D-79104 Freiburg, Germany

Prof. Dr. W. **KRIZ,** Anatomisches Institut der Universität Heidelberg, Im Neuenheimer Feld 307, D-69120 Heidelberg, Germany

Prof. Dr. E. **MARANI**, Leiden University, Department of Physiology, Neuroregulation Group, P.O. Box 9604, 2300 RC Leiden, The Netherlands

Prof. Dr. R. **PUTZ**, Anatomische Anstalt der Universität München, Lehrstuhl Anatomie I, Pettenkoferstr. 11, D-80336 München, Germany

Prof. Dr. Dr. h.c. Y. **SANO,** Department of Anatomy, Kyoto Prefectural University of Medicine, Kawaramachi-Hirokoji, 602 Kyoto, Japan

Prof. Dr. Dr. h.c. T. H. **SCHIEBLER,** Anatomisches Institut der Universität, Koellikerstraße 6, D-97070 Würzburg, Germany

Prof. Dr. K. **ZILLES**, Universität Düsseldorf, Medizinische Einrichtungen, C. u. O. Vogt-Institut, Postfach 101007, D-40001 Düsseldorf, Germany

GPSR Compliance

The European Union's (EU) General Product Safety Regulation (GPSR) is a set of rules that requires consumer products to be safe and our obligations to ensure this.

If you have any concerns about our products, you can contact us on ProductSafety@springernature.com

In case Publisher is established outside the EU, the EU authorized representative is:

Springer Nature Customer Service Center GmbH
Europaplatz 3
69115 Heidelberg, Germany

Batch number: 10704713

Printed by Printforce, the Netherlands